SpringerBriefs in Earth Sciences

More information about this series at http://www.springer.com/series/8897

Swapan Kumar Maity · Ramkrishna Maiti

Sedimentation in the Rupnarayan River

Volume 2: Estuarine Environment of Deposition

 Springer

Swapan Kumar Maity
Department of Geography
Nayagram P.R.M. Government College
Jhargram, West Bengal
India

Ramkrishna Maiti
Department of Geography and Environment
 Management
Vidyasagar University
Paschim Medinipur, West Bengal
India

ISSN 2191-5369 ISSN 2191-5377 (electronic)
SpringerBriefs in Earth Sciences
ISBN 978-3-319-71314-4 ISBN 978-3-319-71315-1 (eBook)
https://doi.org/10.1007/978-3-319-71315-1

Library of Congress Control Number: 2017945254

© The Author(s) 2018
This work is subject to copyright. All rights are reserved by the Publisher, whether the whole or part of the material is concerned, specifically the rights of translation, reprinting, reuse of illustrations, recitation, broadcasting, reproduction on microfilms or in any other physical way, and transmission or information storage and retrieval, electronic adaptation, computer software, or by similar or dissimilar methodology now known or hereafter developed.
The use of general descriptive names, registered names, trademarks, service marks, etc. in this publication does not imply, even in the absence of a specific statement, that such names are exempt from the relevant protective laws and regulations and therefore free for general use.
The publisher, the authors and the editors are safe to assume that the advice and information in this book are believed to be true and accurate at the date of publication. Neither the publisher nor the authors or the editors give a warranty, express or implied, with respect to the material contained herein or for any errors or omissions that may have been made. The publisher remains neutral with regard to jurisdictional claims in published maps and institutional affiliations.

Printed on acid-free paper

This Springer imprint is published by Springer Nature
The registered company is Springer International Publishing AG
The registered company address is: Gewerbestrasse 11, 6330 Cham, Switzerland

Preface

Presently, the sedimentation on river bed is appealing the growing attention of most of the geomorphologists, hydrologists, engineers and planners as it is very important to control the geomorphological, hydrological and ecological characteristics of the river in managing a variety of detrimental problems like the shifting of river course, shortage of water for utilization in different purposes, hindrance of easy discharge of water and upstream flood, navigation difficulties, river bank erosion and loss of settlements and properties, etc. The main aim of the book is to understand and explain the causes, mechanisms and extent of river sedimentation in connection to the seasonal fluctuation of stream energy, environment of sediment deposition, sources of sediments and their distributional pattern.

Available shear stress and critical shear stress during high and low tides in different seasons in the stream have been calculated following DuBoys formula and Shield formula. Environment of sediment deposition is identified by Linear Discriminate Analysis and Bi-variate plotting of sediment grain size parameters. Sources of sediments are understood through identification of mineral composition of sediments by X-ray diffraction technique.

Rapid rate of sedimentation in the studied river is the result of combined interaction of riverine and marine processes. Seasonal fluctuation of available shear stress and sediment transport capacity during high and low tides in connection to grain size of sediments is the main causative and controlling factor of sedimentation. Sediments are deposited in moderate to lesser violent hydrodynamic condition in estuarine environment and are supplied from the upper catchment and the river banks. The result of the work will be extremely supportive and helpful to the engineers, hydrologists, planners and other concerned authorities, working on the aspects of sedimentation and management of associated problems not only in the study area but also in any of the tidal river in the world.

Midnapur, West Bengal, India　　　　　　　　　　　　　　　Swapan Kumar Maity
　　　　　　　　　　　　　　　　　　　　　　　　　　　　　　Ramkrishna Maiti

Acknowledgements

We would like to express our humble regards and thankfulness to some of our students Sanjay, Arindam, Pratap, Subrata, Tarun, Subhasis and Suman for their continuous help and support during the entire research work. Without their untired help and active participation, the completion of our research work could have been impossible. We also like to extend our immense sense of gratitude to Dr. Animesh Majee, Dr. Prasenjit Bhunia, Mr. Sarbeswar Haldar, Mr. Rajkumar Ghorai, Mr. Alakesh Samanta and Sk. Nurul Alam for their assistance during conducting the research, technical support and valuable suggestions.

We convey our appreciation and thanks to Dr. Moumita Moitra Maiti, Assistant Professor, Department of Geography, Raja N.L. Khan Women's College for her immense support throughout the work.

We are highly obliged to the concerned authority of Kolaghat Thermal Power Station, Kolkata Port Trust, Sub-divisional Office at Tamluk and Ghatal, Geological Survey of India, Survey of India for providing us necessary documents and data. We are especially thankful to the Central Research Facility of Indian Institute of Technology (Kharagpur) for helping us in sedimentological analysis.

We convey a lot of thanks to all the members of our family for their encouragement, co-operation and continuous moral and emotional supports.

Midnapur, West Bengal, India

Swapan Kumar Maity
Ramkrishna Maiti

Contents

1	**Introduction**	1
	1.1 Prologue	1
	References	4
2	**Available and Critical Shear Stress for Sediment Entrainment**	5
	2.1 Introduction	5
	2.2 Field Monitoring and Applied Methodology	6
	2.3 Shear Stress at Different Stations	7
	2.3.1 Shear Stress Near Kolaghat	7
	2.3.2 Shear Stress Near Soyadighi	9
	2.3.3 Shear Stress Near Anantapur	13
	2.3.4 Shear Stress Near Pyratungi	17
	2.3.5 Shear Stress Near Dhanipur	20
	2.3.6 Shear Stress Near Geonkhali	20
	2.4 Seasonal Fluctuation of Available Shear Stress Along the Lower Reach	22
	2.5 Seasonal Fluctuation of Critical Shear Stress Along the Lower Reach	30
	2.6 Factors Affecting Sedimentation in the Lower Reach	33
	2.6.1 Sedimentation Due to Deficiency of Available Shear Stress During Low Tide Condition	33
	2.6.2 Sedimentation Due to Sheltering and Packing Effects in Non-uniform Sediments	35
	2.6.3 Sedimentation Due to Effects of Cohesion and Adhesion Forces	36
	2.6.4 Sedimentation Due to Biological Influence	37
	References	37

3	**Environment of Sediment Deposition**		39
	3.1	Introduction	39
	3.2	Materials and Methodology of the Study	40
	3.3	Linear Discriminate Analysis (LDA) of Sediments	41
		3.3.1 Environment of Sediment Deposition at Kolaghat	41
		3.3.2 Environment of Sediment Deposition at Soyadighi	42
		3.3.3 Environment of Sediment Deposition at Anantapur	43
		3.3.4 Environment of Sediment Deposition at Pyratungi	44
		3.3.5 Environment of Sediment Deposition at Dhanipur	45
		3.3.6 Environment of Sediment Deposition at Geonkhali	46
	3.4	Influences of Fluvial and Marine Processes	47
		3.4.1 Spatial and Seasonal Variation of Influences of Fluvial and Marine Processes	48
	3.5	Bi-variate Plot of Mean and Standard Deviation	48
	3.6	Bi-variate Plot of Skewness and Standard Deviation	51
	3.7	Deflection of Marine and Riverine Flow by Coriolis Force	51
	3.8	C-M Plotting to Identify Hydrodynamic Forces Working During Deposition	53
	References		55
4	**Identification of the Sediment Sources Using X-Ray Diffraction (XRD) Technique**		57
	4.1	Introduction	57
	4.2	Significance of the Understanding of Sources of Sediments in the Study Area	58
	4.3	Mineralogy of the Catchment Area of the Rupnarayan River	59
	4.4	Materials and Methodology of the Study	60
		4.4.1 Treatments of Collected Sediment Samples	60
		4.4.2 XRD Analysis of Sediment Samples	62
		4.4.3 Quantification of Different Minerals in Sediments	64
	4.5	Mineralogy of Sediments	67
		4.5.1 Mineralogy of River Bank Sediment Samples	67
		4.5.2 Mineralogy of River Bed Sediment Samples	67
	4.6	Principal Component Analysis (PCA) of Bed Sediment Minerals	69
	4.7	Understanding the Sources of Sediments	71
	References		77
5	**Conclusion**		79
Index			81

Abbreviations

CRF	Central Research Facility
DPM	District Planning Maps
GPS	Global Positioning System
GSI	Geological Survey of India
IIT	Indian Institute of Technology
KPT	Kolkata Port Trust
KTPP	Kolaghat Thermal Power Project
KTPS	Kolaghat Thermal Power Station
LDA	Linear Discriminate Analysis
PCA	Principal Component Analysis
XRD	X-Ray Diffraction

Symbols

D	Median grain size (m)
D	Water depth
R	Hydraulic radius
S	River bed slope
d	Distance between planes of atoms
g	Gravitational acceleration
h	Horizontal distance
θ	Slope
θ	Angle of incidence
ϕ	Sediment grain size unit
λ	X-ray wavelength
ρ	Water density
ρs	Sediment density
μ	Dynamic viscosity
γ	Specific weight of the water
γ_s	Specific weight of sediment
τ_0	Available shear stress
τ_{cr}	Critical shear stress
x_0	Reference grain size fixed at 1 millimetre

List of Figures

Fig. 2.1	Critical shear stress, available shear stress and shear velocity at Kolaghat	9
Fig. 2.2	Critical shear stress, available shear stress and shear velocity at Soyadighi	13
Fig. 2.3	Critical shear stress, available shear stress and shear velocity at Anantapur	17
Fig. 2.4	Critical shear stress, available shear stress and shear velocity at Pyratungi	22
Fig. 2.5	Critical shear stress, available shear stress and shear velocity at Dhanipur	26
Fig. 2.6	Critical shear stress, available shear stress and shear velocity at Geonkhali	30
Fig. 2.7	Surplus and deficiency of stream energy during Pre-monsoon (**a**) and Post-monsoon (**b**)	34
Fig. 2.8	Moving character of non-uniform sediments (Bettes 2008)	36
Fig. 2.9	Influence of biological activity **a** near Pyratungi and **b** near Kolaghat	37
Fig. 3.1	**a** Fluvial and marine influence in pre-monsoon season. **b** Fluvial and marine influence in monsoon season. **c** Fluvial and marine influence in post-monsoon season	49
Fig. 3.2	**a** Bi-variate plot of mean versus standard deviation in pre-monsoon season. **b** Bi-variate plot of mean versus standard deviation in monsoon season. **c** Bi-variate plot of mean versus standard deviation in post-monsoon season	50
Fig. 3.3	**a** Bi-variate plot of standard deviation versus skewness in pre-monsoon season. **b** Bi-variate plot of standard deviation versus skewness in monsoon season. **c** Bi-variate plot of standard deviation versus skewness in post-monsoon season	52

Fig. 3.4	Deflection of marine and riverine flow by Coriolis force (Pethick 1984)	53
Fig. 3.5	More sedimentation towards right bank due to rightward deflection of weaker riverine flow by Coriolis force	53
Fig. 3.6	C-M pattern of sediments	54
Fig. 4.1	Location of sediment samples collected for XRD analysis (Maity and Maiti 2016)	61
Fig. 4.2	Sediment sample collection (**a**), preparation of sediment samples (**b**), prepared sediment samples (**c**) and sample pressed in sample holder (**d**)	61
Fig. 4.3	Sediment added to the machine (**a**) and generation of diffractogram (**b**)	63
Fig. 4.4	Methods of mineral identification from diffractogram	63
Fig. 4.5	Quantity of different minerals in sediments (Maity and Maiti 2016)	69
Fig. 4.6	Spatial distribution of quartz (**a**), oligoclase (**b**), illite (**c**) and goethite (**d**) in the lower reach	74
Fig. 4.7	Spatial distribution of chromite (**a**), sillimanite (**b**), corundum (**c**) and chlorite (**d**) in the lower reach	75
Fig. 4.8	Spatial distribution of chloritoid and mirabilite (**a**), dolomite and epidote (**b**), calcite and staurolite (**c**) and garnet pyrope (**d**) in the lower reach	76

List of Tables

Table 2.1	Shear stress at Kolaghat in Pre-monsoon season	8
Table 2.2	Shear stress at Kolaghat in monsoon season	10
Table 2.3	Shear stress at Kolaghat in post-monsoon season	11
Table 2.4	Shear stress at Soyadighi in pre-monsoon season	12
Table 2.5	Shear stress at Soyadighi in monsoon season	14
Table 2.6	Shear stress at Soyadighi in post-monsoon season	15
Table 2.7	Shear stress at Anantapur in pre-monsoon season	16
Table 2.8	Shear stress at Anantapur in monsoon season	18
Table 2.9	Shear stress at Anantapur in post-monsoon season	19
Table 2.10	Shear stress at Pyratungi in pre-monsoon season	21
Table 2.11	Shear stress at Pyratungi in monsoon season	23
Table 2.12	Shear stress at Pyratungi in post-monsoon season	24
Table 2.13	Shear stress at Dhanipur in pre-monsoon season	25
Table 2.14	Shear stress at Dhanipur in monsoon season	27
Table 2.15	Shear stress at Dhanipur in post-monsoon season	28
Table 2.16	Shear stress at Geonkhali in pre-monsoon season	29
Table 2.17	Shear stress at Geonkhali in monsoon season	31
Table 2.18	Shear stress at Geonkhali in post-monsoon season	32
Table 2.19	Sand-mud ratio in sediments	35
Table 3.1	Result of linear discriminate analysis of sediments at Kolaghat	42
Table 3.2	Result of linear discriminate analysis of sediments at Soyadighi	43
Table 3.3	Result of linear discriminate analysis of sediments at Anantapur	44
Table 3.4	Result of linear discriminate analysis of sediments at Pyratungi	45
Table 3.5	Result of linear discriminate analysis of sediments at Dhanipur	46
Table 3.6	Result of linear discriminate analysis of sediments at Geonkhali	47

Table 4.1	Mineralogy of the lower catchment area (Ghosh and Datta 1974)	60
Table 4.2	Grain size characteristics of bed sediments (Maity and Maiti 2016)	62
Table 4.3	Instrumental conditions of XPERT-PRO diffractometer (Maity and Maiti 2016)	63
Table 4.4	Mineralogy of sediment samples collected from river bed (Maity and Maiti 2016)	65
Table 4.5	Mineralogy of sediment samples collected from river banks (Maity and Maiti 2016)	68
Table 4.6	Statistical significance of minerals distribution (Maity and Maiti 2016)	70
Table 4.7	*Co-relation matrix* of different minerals	72
Table 4.8	Depiction of loadings on principal components	73

Chapter 1
Introduction

Abstract The *causes and mechanisms of sedimentation* in the lower reach of the Rupnarayan River are explained in connection to *stream energy, environment of sediment deposition* and understanding of the *sediment sources* from mineral composition. Seasonal fluctuation of *available and critical shear stress*, cohesiveness of sediments, sheltering of sediment particles, effects of biological activity and organic content are the main factors to control the mechanisms of sedimentation. Environment of deposition is examined and understood by *Linear Discriminate Analysis (LDA)* technique and bi-variate plotting of grain size parameters. Hydrodynamic processes working during the deposition of sediments have been identified by *C-M plotting. X-ray diffraction (XRD)* technique is used to understand the sources of sediments through identification of mineral composition.

Keywords Mechanisms of sedimentation · Available and critical shear stress
Linear Discriminate Analysis (LDA) · Depositional environment
Hydrodynamic processes · Sources of sediments

1.1 Prologue

In the *lower reach* of the *Rupnarayan River* (Fig. 1.1 in Maity and Maiti 2018) continuous *sedimentation*, raising of river bed and development of shoal area create varieties of detrimental problems and impacts on society, economies and the environment including the shifting of river course, shortage of water for utilization in different purposes, hindrance of easy discharge of water and upstream flood, navigation difficulties, river bank erosion and loss of settlements and properties etc. Presently the *problem of sedimentation* is drawing the increasing attention of most of the geomorphologists, hydrologists, engineers and planners for the understanding of the mechanisms of sedimentation, reduction of the sedimentation rate and management of the associated problems. Detailed studies on all the geomorphological and *hydrodynamic* characteristics of the studied river indicate that forms and patterns of the channel, hydraulic characteristics of the stream and the nature and characteristics of tide all play a combined role to control the mechanisms and dimensions of sedimentation (Maity and Maiti 2018). All the cross-profiles (except two, one near Kolaghat and other near Geonkhali) drawn along the lower reach of the studied river are *asymmetrical* in nature leading to the concentration of energy near one bank (deeper section of the cross-section) and shortage of energy towards another bank which invites sedimentation. Sudden *widening of the channel* and *diversion of flow* and associated variation in the distribution of energy across the channel play significant role to control the sedimentation process. Near Kolaghat region, abrupt widening of the channel leads to flow separation, reduction of energy and sediment transport capacity which accelerates the rate of sedimentation during low tide. Again, near Geonkhali region, sudden decrease of channel width (bottle neck shape) hinders the easy discharge of ebb tide water causing the reduction of water velocity and stream energy which causes sedimentation (Maity and Maiti 2018). As the catchment area of the river experiences typical monsoonal type of climate, seasonal concentration of rainfall and associated variation of seasonal *water discharge* is very important in controlling the stream energy, sediment transporting capacity and rate of sedimentation. In dry season, less discharge of water from upstream (850–4160 m^3/sec) causes the reduction of stream energy and sediment transporting capacity. Because of this, all the sediments are not drained freely towards downstream during low tide giving enough opportunity for the deposition of sediments on the river bed. But the situation becomes reverse during monsoon season when voluminous water discharge (3455–9050 m^3/sec) from upstream (due to occurrence of huge rainfall) increases the stream energy and sediment transporting capacity causing the easy draining of sediments towards downstream during low tide and the possibility and rate of sedimentation is reduced (Maity and Maiti 2018). *Tidal asymmetry* is very significant to control the sedimentation mechanism in the lower reach. The duration of high tide is shorter by 2–6 h than that of low tide in the studied river. Thus, during swifter high tide condition the availability of greater energy leads to easy transport of sediment towards upstream. But, the sluggish low tide discharge over longer duration (8–9 h) gives sufficient opportunity for the deposition of sediments on the river bed (Maity and Maiti 2018).

During non-monsoon period, upstream penetration of suspended sediment is more during high tide (3.71×10^7–1.29×10^8 metric tons/year) than that is discharged towards downstream during low tide (2.5×10^7–1.0×10^8 metric tons/year) which accelerates the rate of *sedimentation*. But in monsoon, the transport of suspended sediment during high tide (7.5×10^7–2.45×10^8 metric tons/year) and low tide (7.3×10^7–2.3×10^8 metric tons/year) is almost equal which restricts the *sedimentation* rate (Maity and Maiti 2018). Bed load transport rate varies from 0.1905 to 6.52985 kg/m/sec, 0.5008 to 14.74893 kg/m/sec and 0.2318 to 6.31764 kg/m/sec in pre-monsoon, monsoon and post-monsoon season respectively. In non-monsoon season, the transport of bed load is more during high tide than during low tide, but the transport of bed load is almost equal in both the tidal phases in monsoon season. Texturally, sediments are coarser in monsoon than in pre-monsoon and post-monsoon seasons due to increase of water volume, stream energy and removal of fine sediments in monsoon (Maity and Maiti 2018). In dry season, >60% sediments are moderately to well sorted but in monsoon season 63.85% sediments are poorly to very poorly sorted. Generally, the coarser sediments are negatively skewed and finer sediments are positively skewed. Proportion of sand, silt and clay in sediments ranges between 38–91%, 4–61% and 1–41% respectively. Nearly, 81.33% of the sediments are silty sand, 7.33% are muddy sand and 6% samples are of sandy silt category (Maity and Maiti 2018).

Though the role of geomorphological, hydrodynamic and sediment characteristics of the river in controlling the mechanisms and extent of sedimentation in the lower reach of the Rupnarayan River is clearly explained and understood but it is indispensable to explain the seasonal fluctuation of *available shear stress* during high and low tide in connection to *critical shear stress* and the environmental conditions of sediment deposition for the understanding of the causes and *mechanisms of sedimentation* in a holistic way. Availability of *shear stress* to transport the sediment and its seasonal fluctuation during high tide and low tide is an important factor to control the erosional and depositional behavior of a river. Dey (1999) mentioned that over the loose sedimentary bed, the sediments start to move when the continuous increase of water velocity leads to the increase of bed shear stress (available shear stress) and exceeds a critical value of shear stress. Again, the degree of *cohesiveness of sediments* (depends on the ratio of sand-mud mixture) plays significant role in controlling the rate of erosion and transportation of sediment fractions. Different factors like, *sheltering,* imbrications, armoring, algal cover and variations in sorting also affect the resistance, and in turn the critical shear stress required to entrain the sediment of a particular size (Charlton 2007; Clayton 2010; Mayoral 2011).

Different statistical parameters (*Mean, Median, Sorting, Skewness, Kurtosis* etc.) of sediment grain size depict the *environment of sediment deposition* along with the reflection of hydraulic conditions of transporting medium and depositional basin (McLaren and Bowels 1985; Ghosh and Chatterjee 1994). Nature of sorting of sediments, sign of skewness and the values of kurtosis are also indicative of nature of flow and energy condition of the depositing environment. *Positive skewness* indicates the unidirectional transport of sediments or the deposition of sediments in sheltered *low energy environment* (Brambati 1969). Friedman (1962) suggested that extreme high or low values of *kurtosis* imply that part of the sediment achieved its sorting elsewhere in a high energy environment (Baruah et al. 1997). Generally, coarser sediments are deposited *in high-energy environments* whereas the finer sediments are deposited in *low energy conditions. Linear Discriminate Analysis (LDA)* uses different statistical parameters to explain the environmental conditions of sediment deposition (Sahu 1964). Identification of the *mineralogy of sediments* is an indicative of their sources and the environmental conditions as minerals are mostly deposited depending on the variations of their size, shape, density, specific gravity and degree of solubility etc. Rittenhouse (1943), Friedman (1961) and Blatt et al. (1980) indicated that *mineral composition* of sediments along with their textural characteristics reflects the nature of the region of the sediment source and the depository environment.

In the present study, all the above mentioned factors including *available and critical shear stress, environment of sediment deposition, understanding of the sediment sources by identifying sediment mineralogy* etc. have been taken into consideration to analyze and explain the causes and mechanisms of sedimentation systematically and scientifically as a complex interaction between fluvial and tidal processes in the lower reach of the Rupnarayan River.

References

Baruah J, Kotoky P, Sarma JN (1997) Textural and geochemical study on river sediments: a case study on the Jhanji river, Assam. J Indian Assoc Sedimentol 16:195–206

Blatt H, Middleton G, Murray R (1980) Origin of sedimentary rocks. Prentice-Hall, Englewood Cliffs, NJ

Brambati A (1969) Stratigraphy and sedimentation of Siwaliks of North Eastern India. In: Proceedings of international seminar on intermontane basins: geology and resources, Chiang Mai, Thailand, pp 427–439

Charlton R (2007) Fundamentals of fluvial geomorphology. Routledge, New York, NY, p 234

Clayton J (2010) Local sorting, bend curvature, and particle mobility in meandering gravel bed rivers. Water Res Res. 46. https://doi.org/10.1029/2008WR007669

Dey S (1999) Sediment threshold. Appl Math Model 23:399–417

Friedman GM (1961) Distinction between dune, beach and river sands from their textural characteristics. J Sediment Petrol 31(4):514–529

Friedman GM (1962) On sorting, sorting co-efficient and log—normality of the grain size distribution of sandstones. J Geol 70:737–753

Ghosh SK, Chatterjee BK (1994) Depositional mechanism as revealed from grain size measures of the Palaeoproterozoic Kolhan Siliciclastics. Keonjhar District, Orissa, India. Sediment Geol. 89:181–196

Maity SK, Maiti RK (2018) Sedimentation in the Rupnarayan River: hydrodynamic processes under a tidal system. Springer Briefs in Earth Sciences. Springer, Berlin

Mayoral H (2011) Particle size, critical shear stress, and benthic invertebrate distribution and abundance in a Gravel-bed River of the Southern Appalachians. Geosciences Theses, Paper 31

McLaren P, Bowels SD (1985) The effects of sediment transport on grain size distribution. J Sediment Petrol 55(4):457–470

Rittenhouse G (1943) A visual method of estimating two dimensional sphericity. J Sediment Petrol 13:79–81

Sahu BK (1964) Depositional mechanism from the size analysis of clastic sediments. J Sediment Petrol 34(1):73–83

Chapter 2
Available and Critical Shear Stress for Sediment Entrainment

Abstract *Causes and mechanisms of sedimentation* are explained in connection to the seasonal fluctuation of shear stress. *Available and Critical shear stress* have been calculated following DuBoys and Shield formula. Critical shear stress of sediment entrainment varies from 0.031 to 0.147 N/m^2 in pre-monsoon, 0.041 to 0.169 N/m^2 in monsoon and 0.034 to 0.148 N/m^2 in post-monsoon season. Available shear stress varies from 0.271 to 0.923 N/m^2 in high tide and 0.014 to 0.683 N/m^2 in low tide during pre-monsoon season. In monsoon, it varies from 0.275 to 0.965 N/m^2 and 0.237 to 0.907 N/m^2 during high tide and low tide respectively. It varies from 0.259 to 0.889 N/m^2 and 0.022 to 0.521 N/m^2 during high tide and low tide in post-monsoon. Lack of available energy to transport the sediment during low tide (in dry season) is the main reason behind the rapid sedimentation in this area. Most of the places (>75%) having *deficiency of energy* (available shear stress is lower than critical shear stress), during low tide are characterized by deposition of sediments. Presence of mud above the critical limit (30%) in some of the sediment samples generates the *cohesive property*, restricts sediment entrainment and invites sedimentation. Sheltering of fine grains by coarse grains, biological activity and organic content increase the critical shear stress of sediment entrainment and causes sedimentation.

Keywords Sediment texture · Available and critical shear stress
Cohesive property · Sedimentation

2.1 Introduction

The force exerted by flowing water on the bed or banks of a stream is called *available shear stress* that is directly proportional to water density, gravitational acceleration, hydraulic radius and slope of the water surface (Du Boys 1879). The amount of energy needed to entrain the sediment grain (incipient motion) of a particular size is the *critical shear stress*. Dey (1999) mentioned that if flow velocity continue to increase over the loose sedimentary bed, the sediments start moving

© The Author(s) 2018
S. Kumar Maity and R. Maiti, *Sedimentation in the Rupnarayan River*,
SpringerBriefs in Earth Sciences, https://doi.org/10.1007/978-3-319-71315-1_2

when the bed shear stress (available shear stress) generated by the flow exceeds a critical value of shear stress. The particles move in different ways in a stream as the flow characteristics, size of sediment, fluid and sediment densities and the channel conditions vary spatially and temporally. Hjulstrom (1935) derived an empirical curve relating mean velocity to grain size from experiments with sand-size particles and extrapolating to larger and smaller sizes. The critical shear stress required for the entrainment of individual sediment particles for non-cohesive sedimentary bed in unidirectional channel was described by Shield (1936). After that, large numbers of scientists and researchers have triggered their rescarch to determine critical shear stress of sediment erosion (Ahmad et al. 2011). They understood and explained that the critical shear stress of finer sediment particles are dependent on both particle size and bulk density but in larger particles the critical shear stress is dependent only on particle size (Ahmad et al. 2011). Torfs (1995) indicated that conventional non-cohesive formulation for sand fractions could be applicable for the mixture with less than around 3–15% mud content. Erosion behaviour changes dramatically when a small amount of mud is added to a sand bed (Mitchener and Torfs 1996; Williamson and Ockenden 1992). The formula of critical shear stress separately for *sand and mud mixture* to describe the erosion behavior of sediment was proposed by Van Ledden (2003). Sand-mud mixture will exhibit the *cohesive property* when the proportion of mud reaches 20–30% (all sediment less than 0.05 mm in diameter) and below it the mixture is non-cohesive in nature (Van Rijn 1993; Ahmad et al. 2011). Charlton (2007) and Clayton (2010) mentioned that patterns such as sheltering, imbrications, armoring, and variations in sorting can also affect resistance, and in turn the critical shear stress required to entrain the sediment (Mayoral 2011).

2.2 Field Monitoring and Applied Methodology

The *available and critical shear stresses* at different places in the study area in pre-monsoon, monsoon and post-monsoon seasons have been calculated to understand the nature and characteristics of initiation of motion of sediments and its relation to sedimentation is established. Depth of the river during high and low tide is measured using Echo Sounder. Surface water velocity and the velocity of water at different depths in both the tidal phases are measured by floating method and using Digital current meter respectively in three seasons. Sufficient numbers of water and sediment samples have been collected and tested in the laboratory of Indian Institute of Technology (IIT), Kharagpur to determine the density of water and sediment. Stuff readings have been taken using leveling instrument considering a large number of change points and reduced level of each point is calculated using the *bench mark* of 6.5 m near Kolaghat Thermal Power Station. Slope of the river bed is calculated following Eq. 2.1 in Maity and Maiti (2018). Total 180 sediment samples (60 samples in each season) have been collected from six (6) stations (Kolaghat, Soyadighi, Anantapur, Pyratungi, Dhanipur and Geonkhali) at the lower reach (Fig. 6.1 in Maity and Maiti 2018) to understand the textural characteristics of sediments by *sieving technique*. Median (D_{50}) grain size of sediments is determined by using conventional graphical method of Folk and Ward (1957). The places from which sediment samples have been collected were taken to be fixed (by GPS receiver) and monitored in three seasons to identify the nature of shoaling and scouring in the studied reach. Seasonal changes of river depth in those places become helpful to understand the shoaling and scouring phenomena.

Du Boys (1879) equation is used to calculate the *available shear stress* during high and low tide:

$$\tau_0 = \rho g d s \quad (2.1)$$

where, ρ = the water density (1.00 g/cm^3), g = the gravitational acceleration (9.80665 m/s^2), d = the hydraulic radius (m), and s = the river bed slope.

The *critical shear stress* has been calculated following Shield (1936) formula:

$$\tau_{cr} = Kg(\rho_s - \rho)D \quad (2.2)$$

where, K = the constant (0.045), g = the gravitational acceleration, ρ_s = sediment density [typically 2.65 gm/cm^3 (Knighton 1998)], ρ = water density and D = the median grain size in meter.

2.3 Shear Stress at Different Stations

2.3.1 *Shear Stress Near Kolaghat*

As Kolaghat reach is affected by tidal phenomena, shear stress during swifter and stronger high tide is more than that during slower and weaker low tide. Available shear stress is more in monsoon season than in pre-monsoon and post-monsoon seasons. Available shear stress ranges between 0.397–0.745 N/m^2 and 0.415–0.762 N/m^2 during high tide of pre-monsoon and post-monsoon respectively, while during pre-monsoon and post-monsoon low tide it ranges between 0.117–0.436 N/m^2 and 0.109–0.402 N/m^2 respectively (Tables 2.1, 2.3 and Fig. 2.1). In monsoon high tide, available shear stress varies between 0.538 and 0.821 N/m^2 but during low tide it varies between 0.484 and 0.804 N/m^2 (Table 2.2 and Fig. 2.1). During monsoon, critical shear stress to entrain the sediments varies between 0.113 and 0.150 N/m^2, while in pre-monsoon and post-monsoon season it ranges between 0.101–0.147 N/m^2 and 0.099–0.148 N/m^2 respectively. The places of 1, 4, 5, 6 and 7 in pre-monsoon; 4 and 5 in monsoon and 4, 5, 6, 7 and 8 in post-monsoon are characterized by rapid sedimentation (Tables 2.1, 2.2, 2.3 and Fig. 2.1). The deficit of available shear stress in dry season low tide to transport the sediments is the main cause of sediment deposition (Maity and Maiti 2017). The separation of flow, channel widening, packing of grains and effects of organic mat are also responsible

Table 2.1 Shear stress at Kolaghat in Pre-monsoon season

Sediment sample	Mean grain size (m)	Water depth (m)		Water surface slope		Available shear stress (N/m^2) $\tau_0 = \rho g d s$		Critical shear stress (N/m^2) $\tau_{cr} = Kg(\rho_s - \rho)D$	Remarks
		High tide	Low tide	High tide	Low tide	High tide	Low tide		
1	0.000125	2.64	0.84	0.0000152	0.0000140	0.397	0.117	0.120	Deposition
2	0.000121	4.95	3.15			0.745	0.436	0.117	Erosion
3	0.000137	4.35	2.55			0.655	0.353	0.132	Erosion
4	0.000157	2.95	1.15			0.444	0.159	0.147	Deposition
5	0.000143	2.64	0.85			0.397	0.117	0.125	Deposition
6	0.000140	2.75	0.95			0.414	0.131	0.136	Deposition
7	0.000125	2.67	0.87			0.402	0.118	0.123	Deposition
8	0.000120	3.05	1.25			0.459	0.173	0.115	Erosion
9	0.000105	3.65	1.85			0.549	0.256	0.101	Erosion
10	0.000107	3.47	1.67			0.522	0.231	0.108	Erosion

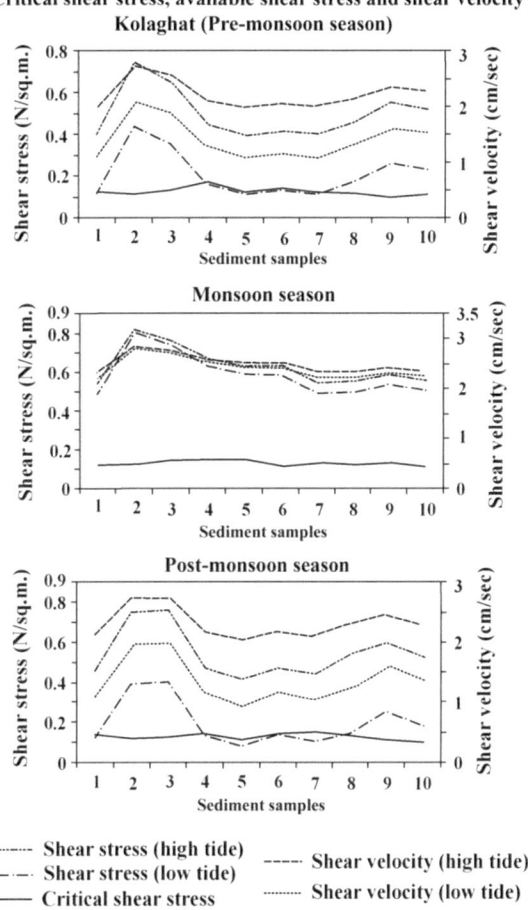

Fig. 2.1 Critical shear stress, available shear stress and shear velocity at Kolaghat. *Source* Field survey and laboratory experiment

for sedimentation. During monsoon season, though the available shear stress is more than critical shear stress, sedimentation happens mainly due to these factors.

2.3.2 Shear Stress Near Soyadighi

During pre-monsoon and post-monsoon high tide, the available shear stress ranges between 0.484–0.843 N/m^2 and 0.466–0.780 N/m^2 respectively, while during pre-monsoon and post-monsoon low tide it ranges between 0.076–0.444 N/m^2 and 0.081–0.391 N/m^2 respectively (Tables 2.4, 2.6 and Fig. 2.2). In monsoonal low tide, the available shear stress varies between 0.480 and 0.855 N/m^2 but during high tide it varies between 0.517 and 0.889 N/m^2 (Table 2.5 and Fig. 2.2). During monsoon, critical shear stress to entrain the sediments varies between 0.107 and

Table 2.2 Shear stress at Kolaghat in monsoon season

Sediment sample	Mean grain size (m)	Water depth (m)		Water surface slope		Available shear stress (N/m^2) $\tau_0 = \rho g d s$		Critical shear stress (N/m^2) $\tau_{cr} = Kg(\rho_s - \rho)D$	Remarks
		High tide	Low tide	High tide	Low tide	High tide	Low tide		
1	0.000125	7.25	5.75	0.0000075	0.0000085	0.538	0.484	0.120	Erosion
2	0.000125	11.05	9.55			0.821	0.804	0.123	Erosion
3	0.000149	10.25	8.75			0.761	0.736	0.143	Erosion
4	0.000158	9.04	7.54			0.672	0.635	0.150	Deposition
5	0.000149	8.55	7.05			0.635	0.593	0.143	Deposition
6	0.000120	8.54	6.90			0.634	0.581	0.115	Erosion
7	0.000140	7.35	5.85			0.546	0.492	0.136	Erosion
8	0.000127	7.45	5.95			0.553	0.501	0.125	Erosion
9	0.000130	7.85	6.35			0.583	0.534	0.134	Erosion
10	0.000115	7.55	6.05			0.561	0.509	0.113	Erosion

2.3 Shear Stress at Different Stations

Table 2.3 Shear stress at Kolaghat in post-monsoon season

Sediment sample	Mean grain size (m)	Water depth (m)		Water surface slope		Available shear stress (N/m²) $\tau_0 = \rho g d s$		Critical shear stress (N/m²) $\tau_{cr} = Kg(\rho_s - \rho)D$	Remarks
		High tide	Low tide	High tide	Low tide	High tide	Low tide		
1	0.000140	3.22	0.92	0.0000142	0.0000130	0.453	0.118	0.136	Erosion
2	0.000121	5.34	3.04			0.751	0.391	0.117	Erosion
3	0.000125	5.42	3.12			0.762	0.402	0.123	Erosion
4	0.000149	3.35	1.05			0.471	0.135	0.142	Deposition
5	0.000119	2.95	0.65			0.415	0.083	0.114	Deposition
6	0.000138	3.35	1.05			0.471	0.135	0.134	Deposition
7	0.000165	3.15	0.85			0.443	0.109	0.148	Deposition
8	0.000135	3.85	1.15			0.541	0.148	0.131	Deposition
9	0.000119	4.25	1.95			0.598	0.251	0.114	Erosion
10	0.000101	3.75	1.45			0.527	0.186	0.099	Erosion

Source Field survey and laboratory experiment

Table 2.4 Shear stress at Soyadighi in pre-monsoon season

Sediment sample	Mean grain size (m)	Water depth (m)		Water surface slope		Available shear stress (N/m^2) $\tau_0 = \rho g d s$		Critical shear stress (N/m^2) $\tau_{cr} = Kg(\rho_s - \rho)D$	Remarks
		High tide	Low tide	High tide	Low tide	High tide	Low tide		
11	0.000125	3.75	0.85	0.0000134	0.0000130	0.498	0.109	0.120	Deposition
12	0.000119	3.92	1.02			0.520	0.113	0.114	Deposition
13	0.000125	4.11	1.21			0.545	0.115	0.123	Deposition
14	0.000136	3.81	0.91			0.505	0.117	0.131	Deposition
15	0.000088	3.45	0.55			0.484	0.076	0.085	Deposition
16	0.000121	4.02	1.12			0.533	0.114	0.116	Deposition
17	0.000125	4.55	1.65			0.604	0.212	0.123	Deposition
18	0.000088	5.29	2.39			0.702	0.308	0.086	Erosion
19	0.000099	6.35	3.45			0.843	0.444	0.096	Erosion
20	0.000095	4.25	1.35			0.564	0.174	0.090	Erosion

Fig. 2.2 Critical shear stress, available shear stress and shear velocity at Soyadighi. *Source* Field survey and laboratory experiment

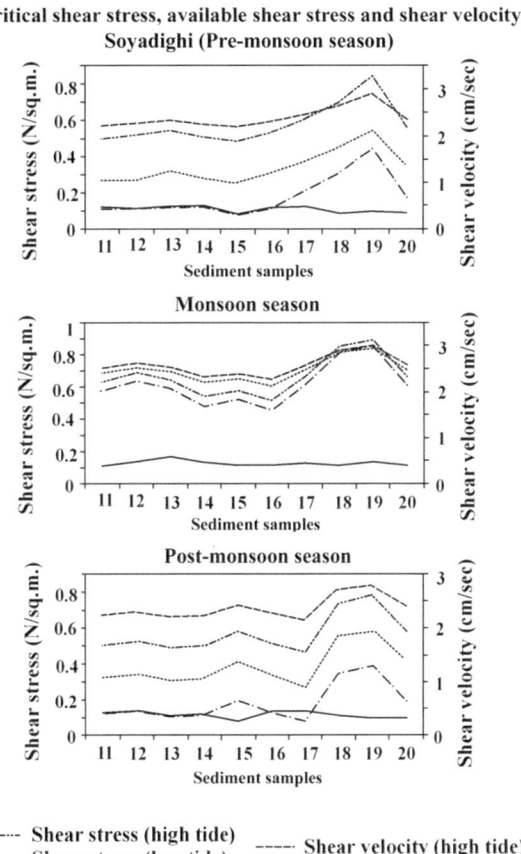

0.169 N/m², while in pre-monsoon and post-monsoon season it ranges between 0.085–0.131 N/m² and 0.078–0.132 N/m² respectively (Maity and Maiti 2017). The sediment samples of locations 11, 12, 13, 14, 15, 16 and 17 in pre-monsoon; 12, 13 and 16 in monsoon and 11, 12, 13, 14, 16 and 17 in post-monsoon season are affected by the mechanisms of sedimentation while other places are characterized by erosion process (Tables 2.4, 2.5, 2.6 and Fig. 2.2).

2.3.3 Shear Stress Near Anantapur

Available shear stress ranges between 0.302–0.686 N/m² and 0.357–0.757 N/m² during high tide of pre-monsoon and post-monsoon respectively, while during pre-monsoon and post-monsoon low tide it ranges between 0.033–0.368 N/m² and

Table 2.5 Shear stress at Soyadighi in monsoon season

Sediment sample	Mean grain size (m)	Water depth (m)		Water surface slope		Available shear stress (N/m²) $\tau_0 = \rho g d s$		Critical shear stress (N/m²) $\tau_{cr} = Kg(\rho_s - \rho)D$	Remarks
		High tide	Low tide	High tide	Low tide	High tide	Low tide		
11	0.000107	6.75	5.75	0.0000094	0.0000101	0.628	0.575	0.107	Erosion
12	0.000139	7.36	6.36			0.685	0.636	0.137	Deposition
13	0.000177	6.90	5.90			0.642	0.590	0.169	Deposition
14	0.000136	5.80	4.80			0.540	0.480	0.132	Erosion
15	0.000112	6.20	5.20			0.577	0.520	0.111	Erosion
16	0.000119	5.55	4.55			0.517	0.455	0.114	Deposition
17	0.000131	7.15	6.15			0.666	0.615	0.126	Erosion
18	0.000119	9.12	8.12			0.849	0.812	0.114	Erosion
19	0.000139	9.55	8.55			0.889	0.855	0.136	Erosion
20	0.000112	7.10	6.10			0.665	0.610	0.112	Erosion

2.3 Shear Stress at Different Stations

Table 2.6 Shear stress at Soyadighi in post-monsoon season

Sediment sample	Mean grain size (m)	Water depth (m)		Water surface slope		Available shear stress (N/m2) $\tau_0 = \rho g d s$		Critical shear stress (N/m2) $\tau_{cr} = Kg(\rho_s - \rho)D$	Remarks
		High tide	Low tide	High tide	Low tide	High tide	Low tide		
11	0.000131	3.94	0.94	0.0000129	0.0000127	0.503	0.118	0.125	Deposition
12	0.000136	4.12	1.12			0.526	0.130	0.132	Deposition
13	0.000107	3.84	0.84			0.490	0.105	0.107	Deposition
14	0.000119	3.91	0.91			0.499	0.115	0.114	Deposition
15	0.000081	4.55	1.55			0.581	0.195	0.078	Erosion
16	0.000136	4.02	1.02			0.514	0.128	0.131	Deposition
17	0.000136	3.65	0.65			0.466	0.081	0.132	Deposition
18	0.000109	5.76	2.76			0.736	0.347	0.108	Erosion
19	0.000100	6.11	3.11			0.780	0.391	0.100	Erosion
20	0.000099	4.55	1.55			0.581	0.195	0.098	Erosion

Source Field survey and laboratory experiment

Table 2.7 Shear stress at Anantapur in pre-monsoon season

Sediment sample	Mean grain size (m)	Water depth (m)		Water surface slope		Available shear stress (N/m^2) $\tau_0 = \rho g d s$		Critical shear stress (N/m^2) $\tau_{cr} = Kg(\rho_s - \rho)D$	Remarks
		High tide	Low tide	High tide	Low tide	High tide	Low tide		
21	0.000085	4.00	0.9	0.0000086	0.0000075	0.340	0.066	0.080	Erosion
22	0.000099	3.55	0.45			0.302	0.033	0.096	Deposition
23	0.000085	4.35	1.15			0.396	0.075	0.080	Deposition
24	0.000085	4.15	1.05			0.353	0.078	0.082	Deposition
25	0.000079	4.12	1.02			0.351	0.074	0.074	Deposition
26	0.000076	6.20	3.10			0.528	0.230	0.072	Erosion
27	0.000053	7.65	4.55			0.651	0.338	0.051	Erosion
28	0.000055	8.06	4.96			0.686	0.368	0.053	Erosion
29	0.000062	6.55	3.55			0.558	0.338	0.060	Erosion
30	0.000069	5.45	2.35			0.464	0.174	0.067	Erosion

Fig. 2.3 Critical shear stress, available shear stress and shear velocity at Anantapur. *Source* Field survey and laboratory experiment

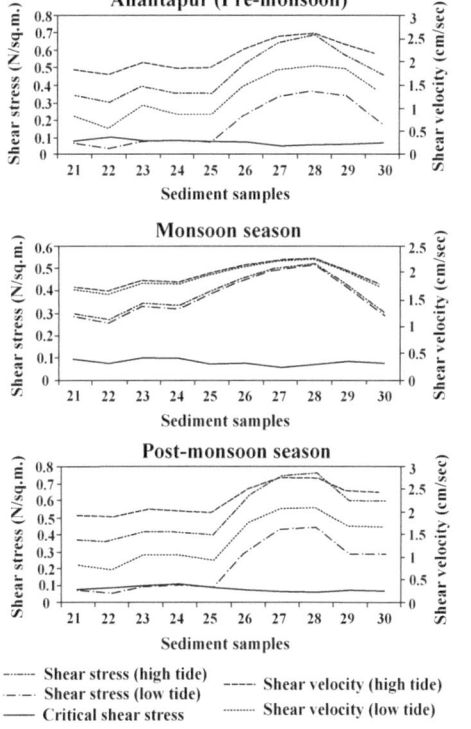

0.054–0.433 N/m² respectively (Tables 2.7, 2.9 and Fig. 2.3). In monsoon high tide, available shear stress varies from 0.275 to 0.518 N/m² but during low tide it varies from 0.256 to 0.514 N/m² (Table 2.8 and Fig. 2.3). During monsoon, critical shear stress to entrain the sediments ranges between 0.060 and 0.102 N/m², while in pre-monsoon and post-monsoon season it ranges between 0.051–0.096 N/m² and 0.060–0.110 N/m² respectively (Maity 2015). Places 22, 23, 24 and 25 in pre-monsoon; 23 and 24 in monsoon and 21, 22, 23, 24 and 25 in post-monsoon season are mostly affected by sediment deposition (Tables 2.7, 2.8, 2.9 and Fig. 2.3).

2.3.4 Shear Stress Near Pyratungi

Available shear stress ranges between 0.345–0.923 N/m² and 0.339–0.776 N/m² during high tide of pre-monsoon and post-monsoon respectively, while during pre-monsoon and post-monsoon low tide it ranges between 0.014–0.682 N/m² and 0.022–0.450 N/m² respectively (Tables 2.10, 2.12 and Fig. 2.4). During monsoon season, the available shear stress in high tide condition varies between 0.416 and

Table 2.8 Shear stress at Anantapur in monsoon season

Sediment sample	Mean grain size (m)	Water depth (m)		Water surface slope		Available shear stress (N/m^2) $\tau_0 = \rho g d s$		Critical shear stress (N/m^2) $\tau_{cr} = Kg(\rho_s - \rho)D$	Remarks
		High tide	Low tide	High tide	Low tide	High tide	Low tide		
21	0.000099	5.05	4.45	0.0000060	0.0000064	0.300	0.282	0.094	Erosion
22	0.000079	4.64	4.04			0.275	0.256	0.074	Erosion
23	0.000100	5.75	5.15			0.342	0.326	0.099	Deposition
24	0.000105	5.62	5.02			0.334	0.318	0.102	Deposition
25	0.000072	6.65	6.05			0.395	0.383	0.070	Erosion
26	0.000079	7.74	7.14			0.460	0.452	0.074	Erosion
27	0.000062	8.48	7.88			0.504	0.499	0.060	Erosion
28	0.000071	8.72	8.12			0.518	0.514	0.069	Erosion
29	0.000085	7.05	6.45			0.419	0.409	0.084	Erosion
30	0.000079	5.15	4.55			0.306	0.288	0.074	Erosion

2.3 Shear Stress at Different Stations

Table 2.9 Shear stress at Anantapur in post-monsoon season

Sediment sample	Mean grain size (m)	Water depth (m)		Water surface slope		Available shear stress (N/m^2) $\tau_0 = \rho g d s$		Critical shear stress (N/m^2) $\tau_{cr} = Kg(\rho_s - \rho)D$	Remarks
		High tide	Low tide	High tide	Low tide	High tide	Low tide		
21	0.000079	3.95	0.75	0.0000095	0.0000092	0.371	0.068	0.075	Deposition
22	0.000091	3.80	0.60			0.357	0.054	0.088	Deposition
23	0.000099	4.35	1.05			0.419	0.095	0.097	Deposition
24	0.000112	4.40	1.20			0.414	0.110	0.110	Deposition
25	0.000088	4.22	1.02			0.397	0.085	0.086	Deposition
26	0.000079	6.65	3.45			0.625	0.314	0.073	Erosion
27	0.000069	7.96	4.76			0.749	0.433	0.067	Erosion
28	0.000062	8.05	4.85			0.757	0.442	0.060	Erosion
29	0.000072	6.30	3.10			0.597	0.282	0.070	Erosion
30	0.000070	6.30	3.10			0.597	0.282	0.068	Erosion

Source Field survey and laboratory experiment

0.965 N/m² but during low tide it varies between 0.313 and 0.907 N/m² (Table 2.11 and Fig. 2.4). Critical shear stress to entrain the sediments varies between 0.054 and 0.095 N/m² in monsoon season, while in pre-monsoon and post-monsoon season it ranges between 0.042–0.086 N/m² and 0.049–0.090 N/m² respectively (Maity and Maiti 2017). Places of 31, 32, 33, 38 and 39 in pre-monsoon; 32, 33 and 39 in monsoon and 31, 32, 33, 37, 38 and 39 in post-monsoon are largely affected by the mechanisms deposition of sediments (Tables 2.10, 2.11, 2.12 and Fig. 2.4).

2.3.5 Shear Stress Near Dhanipur

During high tide of pre-monsoon and post-monsoon, the available shear stress ranges between 0.271–0.449 N/m² and 0.259–0.483 N/m² respectively, while during pre-monsoon and post-monsoon low tide it ranges between 0.041–0.200 N/m² and 0.047–0.261 N/m² respectively (Tables 2.13, 2.15 and Fig. 2.5). In monsoonal flood tide condition, the available shear stress varies between 0.301 and 0.495 N/m² but in low tide condition it varies between 0.237 and 0.437 N/m² (Table 2.14 and Fig. 2.5). During monsoon season, the critical shear stress needed for the transportation of sediments varies between 0.041 and 0.090 N/m², while in pre-monsoon and post-monsoon season it ranges between 0.031–0.068 N/m² and 0.034–0.090 N/m² respectively (Tables 2.13, 2.14, 2.15 and Fig. 2.5) (Maity and Maiti 2017). Sedimentation is happening at the places of 42, 43, 44, 47, 48 and 49 in pre-monsoon; 43, 44 and 48 in monsoon and 43, 44, 47, 48 and 49 in post-monsoon seasons while other places are being affected by erosion mechanism (Tables 2.13, 2.14, 2.15 and Fig. 2.5).

2.3.6 Shear Stress Near Geonkhali

Available shear stress ranges between 0.405–0.892 N/m² and 0.389–0.889 N/m² during high tide of pre-monsoon and post-monsoon respectively, while during pre-monsoon and post-monsoon low tide it ranges between 0.030–0.497 N/m² and 0.036–0.512 N/m² respectively (Tables 2.16, 2.18 and Fig. 2.6). In monsoon season, the available shear stress varies between 0.418–0.839 N/m² and 0.369–0.806 N/m² during high and low tide condition respectively (Table 2.17 and Fig. 2.6). During monsoon, the critical shear stress varies between 0.057 and 0.115 N/m², while in pre-monsoon and post-monsoon season it ranges between 0.031–0.099 N/m² and 0.051–0.096 N/m² respectively (Maity and Maiti 2017). The places of 51, 52, 53, 54, 55 and 56 in pre-monsoon; 53, 54 and 55 in monsoon and 52, 53, 54, 55 and 56 in post-monsoon seasons are characterized by the mechanisms of sedimentation (Tables 2.16, 2.17, 2.18 and Fig. 2.6).

2.3 Shear Stress at Different Stations

Table 2.10 Shear stress at Pyratungi in pre-monsoon season

Sediment sample	Mean grain size (m)	Water depth (m)		Water surface slope		Available shear stress (N/m^2) $\tau_0 = \rho g d s$		Critical shear stress (N/m^2) $\tau_{cr} = Kg(\rho_s - \rho)D$	Remarks
		High tide	Low tide	High tide	Low tide	High tide	Low tide		
31	0.000070	3.35	0.15	0.0000104	0.0000100	0.345	0.014	0.068	Deposition
32	0.000088	3.84	0.64			0.395	0.063	0.085	Deposition
33	0.000067	4.05	0.65			0.417	0.083	0.065	Deposition
34	0.000055	4.25	1.05			0.437	0.104	0.055	Erosion
35	0.000088	5.45	2.25			0.561	0.222	0.086	Erosion
36	0.000059	8.84	5.64			0.910	0.558	0.057	Erosion
37	0.000061	10.1	6.90			0.923	0.683	0.059	Erosion
38	0.000085	3.75	0.55			0.489	0.081	0.084	Deposition
39	0.000085	5.31	2.11			0.547	0.209	0.080	Deposition
40	0.000043	4.28	1.08			0.441	0.107	0.042	Erosion

Fig. 2.4 Critical shear stress, available shear stress and shear velocity at Pyratungi. *Source* Field survey and laboratory experiment

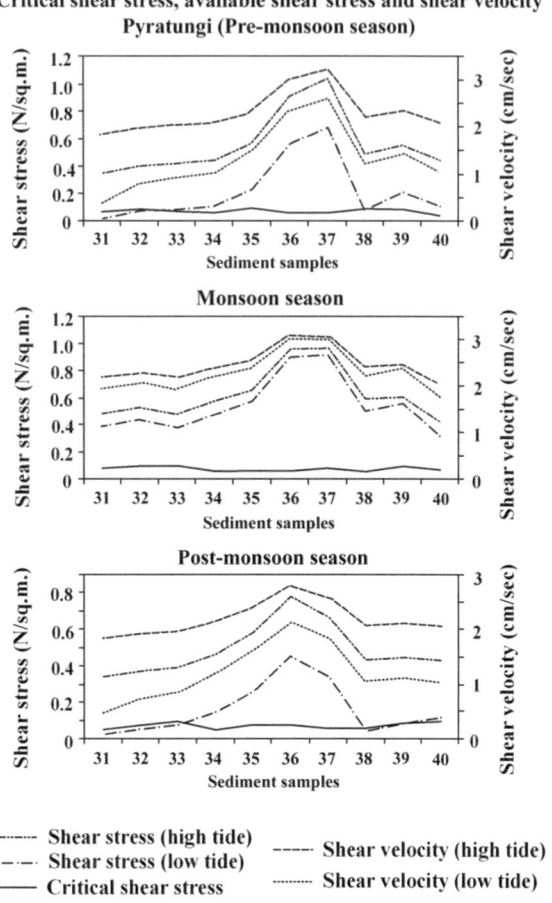

2.4 Seasonal Fluctuation of Available Shear Stress Along the Lower Reach

Available shear stress fluctuates spatially and seasonally depending on the variation of river depth, slope of river bed and water velocity in the area under study. The available shear stress is more in monsoon season than in pre-monsoon and post-monsoon seasons (Tables 2.1, 2.2, 2.3, 2.4, 2.5, 2.6, 2.7, 2.8, 2.9, 2.10, 2.11, 2.12, 2.13, 2.14, 2.15, 2.16, 2.17 and 2.18). This is because of the occurrence of huge rainfall, voluminous terrestrial and riverine discharge, increase of river depth and water velocity in rainy season. In the lower reach, the amount of available shear stress varies from 0.271 to 0.923 N/m² in high tide while it ranges between 0.014 and 0.683 N/m² in low tide during pre-monsoon season (Tables 2.1, 2.2, 2.3, 2.4, 2.5, 2.6, 2.7, 2.8, 2.9, 2.10, 2.11, 2.12, 2.13, 2.14, 2.15, 2.16, 2.17 and 2.18). During monsoon season, the shear stress varies from 0.275 to 0.965 N/m² and

2.4 Seasonal Fluctuation of Available Shear Stress Along the Lower Reach 23

Table 2.11 Shear stress at Pyratungi in monsoon season

Sediment sample	Mean grain size (m)	Water depth (m)		Water surface slope		Available shear stress (N/m^2) $\tau_0 = \rho g d s$		Critical shear stress (N/m^2) $\tau_{cr} = Kg(\rho_s - \rho)D$	Remarks
		High tide	Low tide	High tide	Low tide	High tide	Low tide		
31	0.000079	5.75	4.25	0.0000085	0.0000092	0.484	0.387	0.075	Erosion
32	0.000089	6.26	4.76			0.527	0.433	0.087	Deposition
33	0.000099	5.62	4.12			0.473	0.375	0.095	Deposition
34	0.000058	6.83	5.33			0.575	0.485	0.056	Erosion
35	0.000062	7.78	6.28			0.655	0.572	0.060	Erosion
36	0.000062	11.38	9.88			0.958	0.900	0.057	Erosion
37	0.000081	11.46	9.96			0.965	0.907	0.078	Erosion
38	0.000055	6.97	5.47			0.587	0.498	0.054	Erosion
39	0.000095	7.12	6.12			0.600	0.557	0.092	Deposition
40	0.000071	4.94	3.44			0.416	0.313	0.069	Erosion

Table 2.12 Shear stress at Pyratungi in post-monsoon season

Sediment sample	Mean grain size (m)	Water depth (m)		Water surface slope		Available shear stress (N/m²) $\tau_0 = \rho g d s$		Critical shear stress (N/m²) $\tau_{cr} = Kg(\rho_s - \rho)D$	Remarks
		High tide	Low tide	High tide	Low tide	High tide	Low tide		
31	0.000060	3.43	0.23	0.0000100	0.0000098	0.339	0.022	0.055	Deposition
32	0.000081	3.74	0.54			0.370	0.052	0.076	Deposition
33	0.000095	3.93	0.73			0.389	0.070	0.090	Deposition
34	0.000051	4.65	1.45			0.460	0.140	0.049	Erosion
35	0.000079	5.82	2.62			0.576	0.254	0.074	Erosion
36	0.000081	7.84	4.64			0.776	0.450	0.076	Erosion
37	0.000055	6.73	3.53			0.667	0.342	0.054	Deposition
38	0.000053	3.55	0.35			0.431	0.044	0.052	Deposition
39	0.000085	4.06	0.86			0.442	0.079	0.080	Deposition
40	0.000095	4.32	1.12			0.428	0.108	0.090	Erosion

Source Field survey and laboratory experiment

2.4 Seasonal Fluctuation of Available Shear Stress Along the Lower Reach

Table 2.13 Shear stress at Dhanipur in pre-monsoon season

Sediment sample	Mean grain size (m)	Water depth (m)		Water surface slope		Available shear stress (N/m^2) $\tau_0 = \rho g d s$		Critical shear stress (N/m^2) $\tau_{cr} = Kg(\rho_s - \rho)D$	Remarks
		High tide	Low tide	High tide	Low tide	High tide	Low tide		
41	0.000031	5.06	1.56	0.0000066	0.0000065	0.330	0.100	0.031	Erosion
42	0.000055	4.25	0.75			0.278	0.048	0.053	Deposition
43	0.000061	4.44	0.94			0.290	0.059	0.059	Deposition
44	0.000068	4.38	0.88			0.286	0.056	0.066	Deposition
45	0.000035	6.87	3.37			0.449	0.200	0.033	Erosion
46	0.000037	6.68	3.18			0.436	0.189	0.036	Erosion
47	0.000070	4.15	0.65			0.271	0.041	0.068	Deposition
48	0.000044	4.22	0.72			0.275	0.042	0.044	Deposition
49	0.000061	4.41	0.91			0.288	0.060	0.059	Deposition
50	0.000040	4.52	1.02			0.295	0.065	0.038	Erosion

Fig. 2.5 Critical shear stress, available shear stress and shear velocity at Dhanipur. *Source* Field survey and laboratory experiment

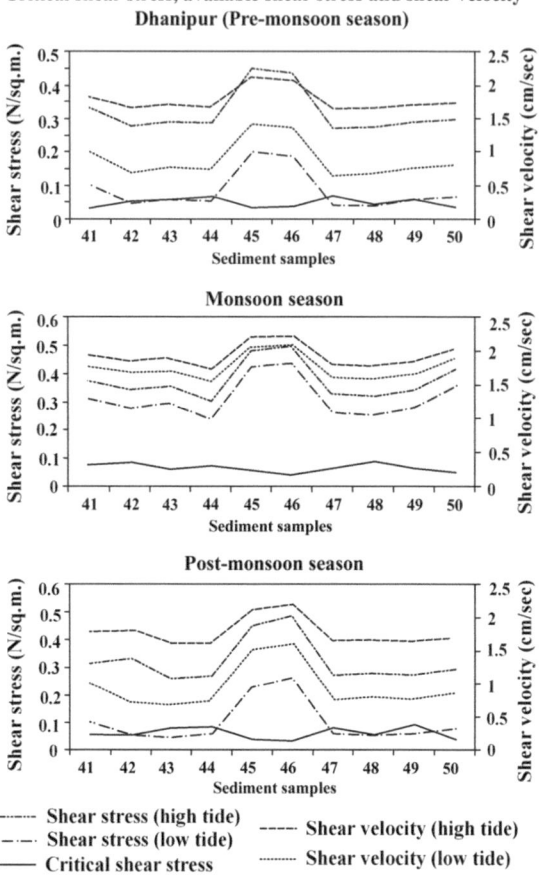

0.237 N/m2 to 0.907 N/m² during high and low tide respectively. It varies from 0.259 to 0.889 N/m² and 0.022 to 0.521 N/m² during high and low tide condition in post-monsoon season (Tables 2.1, 2.2, 2.3, 2.4, 2.5, 2.6, 2.7, 2.8, 2.9, 2.10, 2.11, 2.12, 2.13, 2.14, 2.15, 2.16, 2.17 and 2.18). It is observed that, the available shear stress is higher along the thalweg position of the channel where the river depth and water velocity becomes maximum. Available shear stress during high tide is more than that of in low tide in all the seasons. As the studied reach is dominated by the tidal phenomena, during stronger and shorter high tide (3–4 h) the available shear stress to transport the sediment is increasesd but weaker low tide for longer duration (8–9 h) causes the reduction of available shear stress and sediment transport capacity leading to sedimentation (Tables 2.1, 2.2, 2.3, 2.4, 2.5, 2.6, 2.7, 2.8, 2.9, 2.10, 2.11, 2.12, 2.13, 2.14, 2.15, 2.16, 2.17 and 2.18) (Maity and Maiti 2017).

2.4 Seasonal Fluctuation of Available Shear Stress Along the Lower Reach

Table 2.14 Shear stress at Dhanipur in monsoon season

Sediment sample	Mean grain size (m)	Water depth (m)		Water surface slope		Available shear stress (N/m^2) $\tau_0 = \rho g d s$		Critical shear stress (N/m^2) $\tau_{cr} = Kg(\rho_s - \rho)D$	Remarks
		High tide	Low tide	High tide	Low tide	High tide	Low tide		
41	0.000081	6.26	5.06	0.0000060	0.0000062	0.372	0.311	0.078	Erosion
42	0.000085	5.75	4.55			0.342	0.279	0.084	Erosion
43	0.000064	5.96	4.76			0.354	0.292	0.062	Deposition
44	0.000079	5.06	3.86			0.301	0.237	0.073	Deposition
45	0.000058	8.08	6.88			0.480	0.422	0.056	Erosion
46	0.000042	8.32	7.12			0.495	0.437	0.041	Erosion
47	0.000067	5.47	4.27			0.325	0.262	0.065	Erosion
48	0.000095	5.35	4.15			0.318	0.255	0.090	Deposition
49	0.000067	5.75	4.55			0.342	0.279	0.065	Erosion
50	0.000055	6.94	5.74			0.412	0.353	0.054	Erosion

Table 2.15 Shear stress at Dhanipur in post-monsoon season

Sediment sample	Mean grain size (m)	Water depth (m)		Water surface slope		Available shear stress (N/m^2) $\tau_0 = \rho g d s$		Critical shear stress (N/m^2) $\tau_{cr} = Kg(\rho_s - \rho)D$	Remarks
		High tide	Low tide	High tide	Low tide	High tide	Low tide		
41	0.000053	4.82	1.62	0.0000066	0.0000063	0.315	0.101	0.052	Erosion
42	0.000055	5.05	0.85			0.330	0.053	0.055	Erosion
43	0.000081	3.96	0.76			0.259	0.047	0.079	Deposition
44	0.000085	4.11	0.91			0.267	0.057	0.083	Deposition
45	0.000039	6.87	3.67			0.449	0.229	0.037	Erosion
46	0.000036	7.38	4.18			0.483	0.261	0.034	Erosion
47	0.000085	4.15	0.95			0.271	0.059	0.080	Deposition
48	0.000056	4.12	0.92			0.276	0.055	0.056	Deposition
49	0.000095	4.16	0.96			0.272	0.060	0.090	Deposition
50	0.000040	4.43	1.23			0.290	0.077	0.039	Erosion

Source Field survey and laboratory experiment

2.4 Seasonal Fluctuation of Available Shear Stress Along the Lower Reach

Table 2.16 Shear stress at Geonkhali in pre-monsoon season

Sediment sample	Mean grain size (m)	Water depth (m)		Water surface slope		Available shear stress (N/m^2) $\tau_0 = \rho g d s$		Critical shear stress (N/m^2) $\tau_{cr} = Kg(\rho_s - \rho)D$	Remarks
		High tide	Low tide	High tide	Low tide	High tide	Low tide		
51	0.000031	4.35	0.35	0.0000094	0.0000090	0.405	0.030	0.031	Deposition
52	0.000095	4.54	0.54			0.423	0.048	0.091	Deposition
53	0.000085	4.72	0.72			0.439	0.064	0.084	Deposition
54	0.000044	5.16	1.16			0.667	0.103	0.044	Deposition
55	0.000048	7.44	3.44			0.879	0.236	0.046	Deposition
56	0.000055	8.86	4.86			0.825	0.433	0.053	Deposition
57	0.000085	9.58	5.58			0.892	0.497	0.084	Erosion
58	0.000038	6.10	2.10			0.568	0.187	0.037	Erosion
59	0.000101	5.55	1.55			0.517	0.138	0.099	Erosion
60	0.000044	4.69	0.69			0.437	0.061	0.043	Erosion

Fig. 2.6 Critical shear stress, available shear stress and shear velocity at Geonkhali. *Source* Field survey and laboratory experiment

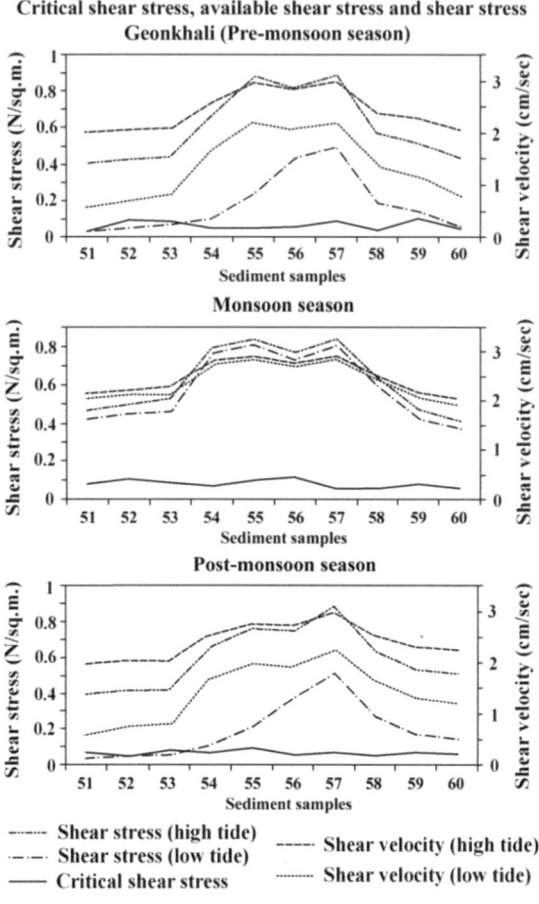

2.5 Seasonal Fluctuation of Critical Shear Stress Along the Lower Reach

The *critical shear stress*, needed for the entrainment of sediments varies spatially and seasonally depending on sediment grain size. It is observed that, the *critical shear stress* continue to increase from downstream to upstream in the lower reach (Tables 2.1, 2.2, 2.3, 2.4, 2.5, 2.6, 2.7, 2.8, 2.9, 2.10, 2.11, 2.12, 2.13, 2.14, 2.15, 2.16, 2.17 and 2.18), though some exceptions are found in few places. Though there is no conspicuous trend of the distribution of sediment grain size towards upstream or downstream (hapazard distribution of sediment grains), yet sediments are slightly finer towards downstream. That's why the critical shear stress decreases gradually towards downstream, because amount of critical shear stress directly depends on grain size of the sediments. Critical shear stress of sediment entrainment is more during monsoon season than in pre-monsoon and post-monsoon seasons

2.5 Seasonal Fluctuation of Critical Shear Stress Along the Lower Reach

Table 2.17 Shear stress at Geonkhali in monsoon season

Sediment sample	Mean grain size (m)	Water depth (m)		Water surface slope		Available shear stress (N/m^2) $\tau_0 = \rho g d s$		Critical shear stress (N/m^2) $\tau_{cr} = Kg(\rho_s - \rho)D$	Remarks
		High tide	Low tide	High tide	Low tide	High tide	Low tide		
51	0.000088	5.96	5.16	0.0000079	0.0000082	0.466	0.419	0.085	Erosion
52	0.000112	6.35	5.55			0.497	0.451	0.111	Erosion
53	0.000090	6.77	5.97			0.530	0.458	0.088	Deposition
54	0.000070	10.15	9.35			0.794	0.760	0.068	Deposition
55	0.000103	10.68	9.88			0.836	0.803	0.101	Deposition
56	0.000119	9.78	8.98			0.765	0.729	0.115	Erosion
57	0.000063	10.72	9.92			0.839	0.806	0.058	Erosion
58	0.000062	8.13	7.33			0.636	0.595	0.057	Erosion
59	0.000079	6.03	5.23			0.472	0.425	0.075	Erosion
60	0.000068	5.34	4.54			0.418	0.369	0.066	Erosion

Table 2.18 Shear stress at Geonkhali in post-monsoon season

Sediment sample	Mean grain size (m)	Water depth (m)		Water surface slope		Available shear stress (N/m^2) $\tau_0 = \rho g d s$		Critical shear stress (N/m^2) $\tau_{cr} = Kg(\rho_s - \rho)D$	Remarks
		High tide	Low tide	High tide	Low tide	High tide	Low tide		
51	0.000070	4.42	0.42	0.0000091	0.0000088	0.398	0.036	0.068	Erosion
52	0.000053	4.65	0.65			0.419	0.050	0.051	Deposition
53	0.000085	4.66	0.66			0.420	0.057	0.084	Deposition
54	0.000071	5.28	1.28			0.656	0.111	0.069	Deposition
55	0.000099	6.44	2.44			0.761	0.212	0.096	Deposition
56	0.000055	8.27	4.27			0.745	0.372	0.054	Deposition
57	0.000071	9.87	5.87			0.889	0.512	0.069	Erosion
58	0.000053	7.10	3.10			0.640	0.270	0.051	Erosion
59	0.000070	5.95	1.95			0.536	0.170	0.068	Erosion
60	0.000062	5.69	1.69			0.513	0.147	0.060	Erosion

Source Field survey and laboratory experiment

2.5 Seasonal Fluctuation of Critical Shear Stress Along the Lower Reach

(Tables 2.1, 2.2, 2.3, 2.4, 2.5, 2.6, 2.7, 2.8, 2.9, 2.10, 2.11, 2.12, 2.13, 2.14, 2.15, 2.16, 2.17 and 2.18). During monsoon period, sample supply of water from upstream and the increase of available shear stress causes easy entrainment of fine sediments and the sediments become coarser (sand fraction increase) than non-monsoon period (pre-monsoon and post-monsoon) (Friedman 1961). As the coarser sediments require more shear stress for entrainment, the critical shear stress in monsoon season increases. In the study area, the amount of critical shear stress ranges between 0.031 and 0.147 N/m^2 (Median grain size 0.000031–0.000157 m), 0.041 and 0.169 N/m^2 (Median grain size 0.000042–0.000177 m) and 0.034 and 0.148 N/m^2 (Median grain size 0.000036–0.000165 m) during pre-monsoon, monsoon and post-monsoon seasons respectively (Tables 2.1, 2.2, 2.3, 2.4, 2.5, 2.6, 2.7, 2.8, 2.9, 2.10, 2.11, 2.12, 2.13, 2.14, 2.15, 2.16, 2.17 and 2.18) (Maity and Maiti 2017).

2.6 Factors Affecting Sedimentation in the Lower Reach

2.6.1 Sedimentation Due to Deficiency of Available Shear Stress During Low Tide Condition

The deficiency of stream energy and reduction of sediment transporting capacity in low tide condition is the main reason behind the rapid sedimentation in the lower reach of the studied river (Maity and Maiti 2017). Sediment deposition is the result of the interaction between available shear stress and the critical shear stress required for the entrainment of sediment of a particular size. Shear stress, available for the entrainment of sediments less than the critical shear stress hinders the easy removal of sediments, leading to rapid *sedimentation, development and expansion of shoal area* (Morisawa 1985). If the available shear stress of a particular place in the river is greater than the critical shear stress then it is termed as *surplus of energy* but if available shear stress is less than critical value then it is known as the *deficiency of energy*. Most of the places, in the lower reach having *deficiency of energy* are affected by the mechanisms of deposition of sediments (Tables 2.1, 2.2, 2.3, 2.4, 2.5, 2.6, 2.7, 2.8, 2.9, 2.10, 2.11, 2.12, 2.13, 2.14, 2.15, 2.16, 2.17 and 2.18). The places 1, 4, 5, 6 and 7 at Kolaghat; 11, 12, 13, 14, 16 at Soyadighi; 21, 22, 23, 24, 25 at Anantapur, 31, 32, 38, 39 at Pyratungi; 42, 43, 44, 47, 48 at Dhanipur; 51, 52 and 53 at Geonkhali are characterized by *deficiency of energy* during low tide in dry season (Fig. 2.7a, b). The spatial distribution of shoaled up area represents that all the above mentioned places are facing the problem of rapid sedimentation (Fig. 6.1 in Maity and Maiti 2018). High tide available shear stress is more (high tide is stronger and faster) than critical shear stress of a particular grain size (Tables 2.1, 2.2, 2.3, 2.4, 2.5, 2.6, 2.7, 2.8, 2.9, 2.10, 2.11, 2.12, 2.13, 2.14, 2.15, 2.16, 2.17 and 2.18), which causes easy landward transportation of sediments but available shear stress during low tide, at some places is lower (low tide is slower and weaker) than

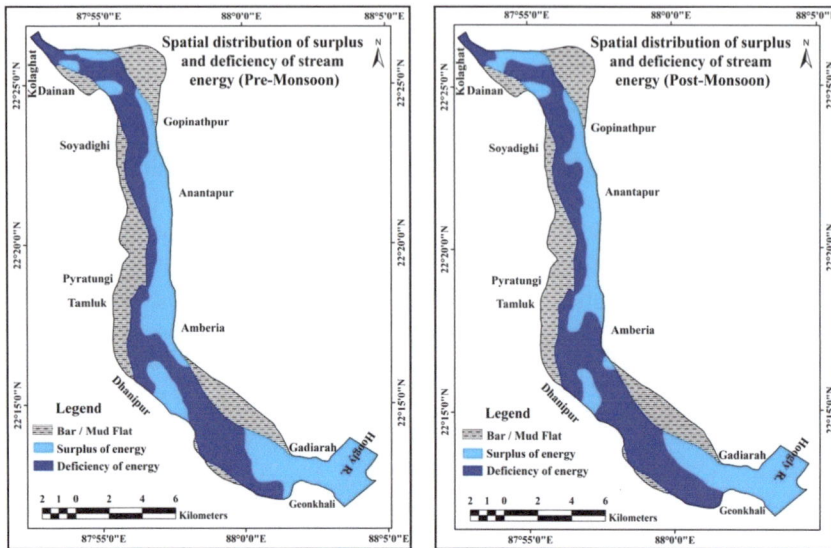

Fig. 2.7 Surplus and deficiency of stream energy during Pre-monsoon (**a**) and Post-monsoon (**b**). *Source* Field survey and laboratory experiment) (Maity and Maiti 2017)

critical shear stress (Tables 2.1, 2.2, 2.3, 2.4, 2.5, 2.6, 2.7, 2.8, 2.9, 2.10, 2.11, 2.12, 2.13, 2.14, 2.15, 2.16, 2.17 and 2.18) which hinders the easy transportation of sediments towards downstream and accelerates the rate of sedimentation in the studied reach (Maity and Maiti 2017).

During monsoon season, occurrence of huge rainfall and *voluminous supply* of *water* from upstream area increases the water velocity during low tide. Due to this reason the available shear stress is almost equal in both the tidal phases during monsoon season, i.e., the sediment transport capacity towards upstream and downstream in both the tidal phases is equal. Because of this the possibility of sediment deposition is minimum and the rate of sedimentation is insignificant in this season (Maity and Maiti 2017). But in pre-monsoon and post-monsoon season, the paucity of rainfall and insufficient supply of water from upper catchment area causes the reduction of water velocity and available shear sress for the entrainment of the sediments, causing a large variation of available shear stress during high and low tide conditions. This deviation of available shear stress in connection to critical shear stress, causes the fluctuation of sediment transport capacity during high and low tide and because of this the rate of sedimentation is high in non-monsoon season than that of in monsoon season (Table 1.2 in Maity and Maiti 2018).

2.6.2 Sedimentation Due to Sheltering and Packing Effects in Non-uniform Sediments

In a number of sediment samples (samples 17 at Soyadighi; 42 and 49 at Dhanipur etc. in pre-monsoon; 4 and 5 at Kolaghat; 23 and 24 at Anantapur; 43, 44 and 48 at Dhanipur etc. in monsoon and 6 and 8 at Kolaghat; 54, 55 and 56 at Geonkhali etc. in post-monsoon season) available shear stress during low tide is more than critical shear stress but sedimentation is observed (Tables 2.1, 2.2, 2.3, 2.4, 2.5, 2.6, 2.7, 2.8, 2.9, 2.10, 2.11, 2.12, 2.13, 2.14, 2.15, 2.16, 2.17 and 2.18). The main reason behind this is the working of the factors like sheltering, imbrications, packing of grains, grain-fabric effects, adhesion forces and organic mats (Charlton 2007; Clayton 2010; Mayoral 2011). In the area under study, sand (38–91%), slit and clay (9–62%) are the main constituents of the sediments (Table 2.19). If sediment is composed of different grain sizes (*non-uniform sediments*), the initiation of motion criteria indicates that the median grain size of sediment will be in motion but actually the coarser fractions of sediment will not move (Fig. 2.8) (Bettes 2008). These coarser fractions of sediments have greater possibility to be deposited on the river bed (Fig. 2.8). Moreover, very fine grained sand, silt and clay particles are sheltered by moderately coarse and coarse sand particles (Wiberg and Smith 1987; Mayoral 2011), which increases the critical shear stress of sediment entrainment and restrict the sediment erosion rate. Hickin (1995) mentioned that sheltered particles need larger force to be entrained than that of same sized grains simply rested on the river bed surface which is fully exposed to the force of the flow. Again, in non-uniform sediments the interstitial spaces stuck between moderately coarse and coarse sand are filled by the fine grained silt and clay particles, resulting the good packing of sediment grains. This mechanism supply additional friction sharing each silt and clay particles in the place surrounded by sand particles (Hickin 1995). Packed mixed-sized sediments require larger force to be entrained than that required to move the same-sized single sediment grains, which resist the easy entrainment of sediments and accelerates the sedimentation rate.

Table 2.19 Sand-mud ratio in sediments

Locations	Sand-mud (silt and clay) proportion (%)					
	Pre-monsoon		Monsoon		Post-monsoon	
	Sand	Mud	Sand	Mud	Sand	Mud
Kolaghat	68–91	9–32	76–91	9–14	71–86	14–29
Soyadighi	60–78	22–40	70–86	14–30	70–84	16–30
Anantapur	45–78	22–55	59–86	14–41	55–78	22–45
Pyratungi	54–79	21–46	73–87	13–27	56–85	15–44
Dhanipur	38–76	24–62	52–84	16–48	45–75	25–55
Geonkhali	46–78	22–54	61–87	13–39	49–74	26–51

Source Field survey and laboratory experiment

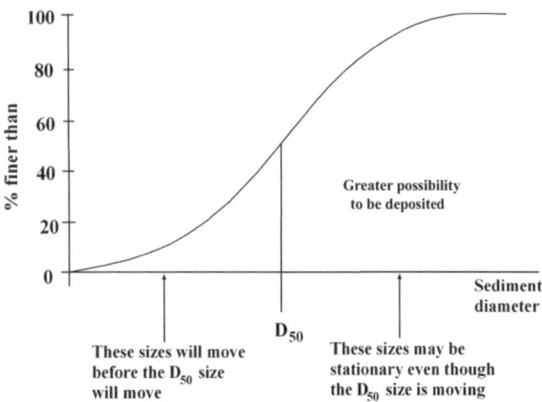

Fig. 2.8 Moving character of non-uniform sediments (Bettes 2008)

2.6.3 Sedimentation Due to Effects of Cohesion and Adhesion Forces

Cohesion and *adhesion force* is the effect of surface-tension of the fine sediment particles and the electrochemical attraction of clay like particles (Hickin 1995). The mechanism of sedimentation in the lower reach is largely affected by cohesion and adhesion forces exerted by the mud particles present in sediments. The degree of cohesiveness of sediments depends on the quantity of mud (silt and clay) present in sediments. Cohesive river bed creates a coherent mass due to electrochemical interactions between the sediment particles and the bed *(adhesion force)* and has certain shear strength before deformation. In this condition, the interactions between sediment particles control the nature of erosion, and the size and weight of sand particles becomes less important (Van Ledden 2003).

The study of Van Rijn (1993) indicated that, when the mud fraction reaches 20–30% then the *sand and mud mixture* will tend to show the *cohesive property* and below this limit the sediment is basically non-cohesive. Mitchener and Torfs (1996) distinguished the non-cohesive sediment from cohesive sediment depending on the mud content being only 3–15%. Van Ledden (2003) mentioned that the cohesive property of sediments is mainly influenced by the critical clay content (5–10%) to generate the cohesiveness. In some of the collected sediment samples in the study area, the proportion of mud (silt and clay) is above the critical limit which generates the *cohesive property*, restricts the sediment entrainment and invites sedimentation. The proportion of mud in sediments is more in dry season (pre-monsoon and post-monsoon) than that of in monsoon season (Table 2.19). Due to this reason, the influence of adhesion force is more in dry season compared to monsoon season. It causes the limited entrainment of silt and clay particles and accelerates the rate of sedimentation in dry season (Table 1.2 in Maity and Maiti 2018). Towards the downstream section of the lower reach of the Rupnarayan River (Pyratungi, Dhanipur and Geonkhali) the proportion of mud is comparatively more which

2.6 Factors Affecting Sedimentation in the Lower Reach

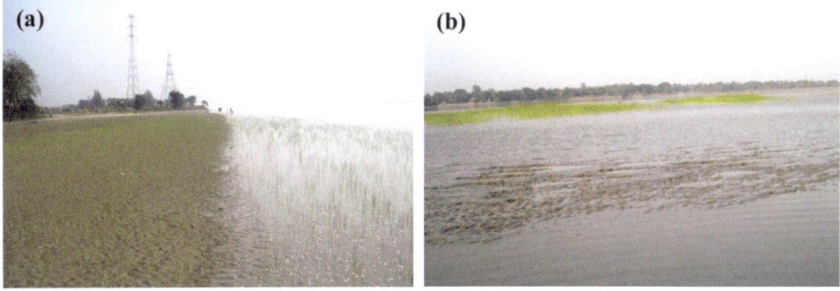

Fig. 2.9 Influence of biological activity **a** near Pyratungi and **b** near Kolaghat

increases the cohesiveness of sediments and restricts their entrainment. Because of this the rate of sedimentation is high towards the lower section of the lower reach (Tables 1.1 and 1.2 in Maity and Maiti 2018). The places, where the proportion of mud is above 30% (14 samples in pre-monsoon, 6 samples in monsoon and 11 samples in post-monsoon), except 6 samples in pre-monsoon and 2 in post-monsoon season all are affected by *the mechanism of sedimentation*.

2.6.4 Sedimentation Due to Biological Influence

The amount of energy required for the initiation of motion of sediment particles on the river bed is largely influenced by biological activities. Paterson et al. (1990) stated that the influence of *biological activity* and *organic content* increase the critical shear stress of sediment entrainment and it becomes more active mainly in dry season. During non-monsoon season when the terrestrial discharge is very meager, *layer of algae and biological content* is developed on river bed which creates a bonding between clay particles in muddy sediments and more critical shear stress is required for the erosion of these sediments. If the voluminous terrestrial discharge during rainy season is unable to disintegrate the surficial algal cover, these sediments are never eroded and continuous deposition of sediments leads to the expansion of shoaled up area (Fig. 2.9) (Maity and Maiti 2017).

References

Ahmad MF, Dong P, Mamat M, Wan Nik WB, Mohd MH (2011) The critical shear stresses for sand and mud mixture. Appl Math Sci 5(2):53–71
Bettes R (2008) Sediment transport and alluvial resistance in rivers. R&D Technical Report W5i 609, Environment Agency, Rio House, Waterside Drive
Charlton R (2007) Fundamentals of fluvial geomorphology. Routledge, New York, NY, p 234

Clayton J (2010) Local sorting, bend curvature, and particle mobility in meandering gravel bed rivers. Water Resour Res. 46. https://doi.org/10.1029/2008WR007669

Dey S (1999) Sediment threshold. Appl Math Model 23:399–417

Du Boys P (1879) Let Rhone et les Rivieres a Lit Affouillable. Annales des Ponts et Chaussees, Series 5(18):141–195

Folk RL, Ward MC (1957) Brazos River bar (Texas): a study in the significance of grain size parameters. J Sediment Petrol 27(1):3–27

Friedman GM (1961) Distinction between dune, beach and river sands from their textural characteristics. J Sediment Petrol 31(4):514–529

Hickin EJ (1995) River geomorphology. Wiley, New York

Hjulstrom F (1935) Studies of the morphological activity of rivers as illustrated by the Rivers Fyris. Bull Geol Inst 25:221–527 University of Uppsala

Knighton D (1998) Fluvial forms and processes: a new perspective. Arnold, London, p 383

Maity SK (2015) Cognition of interworking of processes leading to sedimentation at lower reach of the Rupnarayan River, West Bengal, India. Dissertation, Vidyasagar University, West Bengal, India

Maity SK, Maiti RK (2017) Sedimentation under variable shear stress at lower reach of the Rupnarayan River, West Bengal, India. Water Sci 31:67–92

Maity SK, Maiti RK (2018) Sedimentation in the Rupnarayan River: hydrodynamic processes under a tidal system. Springer Briefs in Earth Sciences. Springer, Berlin

Mayoral H (2011) Particle Size, critical shear stress, and benthic invertebrate distribution and abundance in a Gravel-bed River of the Southern Appalachians. Geosciences Theses. Paper 31

Mitchener H, Torfs H (1996) Erosion of mud/sand mixtures. J Coast Eng 29:1–25

Morisawa M (1985) Rivers: forms and process. Longman Inc, New York

Paterson DM, Crawford RM, Little et al (1990) Sub-aerial exposure and changes in the stability of intertidal estuarine sediments. J Estuar Coast Shelf Sci. 30:541–546

Shield ND (1936) Anwendung der ahnlickeit Mechanik under Turbulenzforschung auf die Geschiebelerwegung. Mitt. Preoss Versuchanstalt fur Wasserbau und Schiffbau, p 26

Torfs H (1995) Erosion of sand/sand mixtures. Dissertation, Catholic University of Leuven, Leuven, Belgium

Van Ledden M (2003) Sand and mud segregation. Dissertation, Delft University of Technology, Delft

Van Rijn LC (1993) Principles of sediment transport in rivers, estuaries and coastal areas. Aqua Publications, The Netherlands

Wiberg PL, Smith JD (1987) Calculations of the critical shear stress for motion of uniform and heterogeneous sediments. Water Resour Res 23(8):1471–1480. https://doi.org/10.1029/WR023i008p01471

Williamson HJ, Ockenden MC (1992) Tidal transport of mud/sand mixtures. Laboratory Tests HR Wallingford, Report SR, p 257

Chapter 3
Environment of Sediment Deposition

Abstract Grain size of sediments relates to the physical characteristics of the *depositional environments*. The distribution of sediment grain size is affected by the variations of wave energy and turbulent conditions of depositing environment. *Linear Discriminate Analysis* (*LDA*) technique and bi-variate plotting of grain size parameters are used to explain and understand the environment of sediment deposition. *Hydrodynamic processes* working during the deposition of sediments have been identified by *C-M plotting*. Sedimentation, in this reach is the result of the spatial and seasonal variation of the interaction between fluvial and marine processes. In non-monsoon season more than 60% of the sediment samples fall under *marine environment*, but in monsoon season more than 65% of the sediment samples fall under *riverine environment*. Nearly, 58% of the sediments are deposited under fluvial action and 42% samples are deposited by turbidity action mainly in low to moderate energy condition. Deflection of sluggish riverine discharge towards right by *Coriolis force* causes more sedimentation towards right bank of the river mainly in non-monsoon season. About 88% sediments are transported by suspension with rolling and graded suspension and are deposited in moderate to lesser violent hydrodynamic condition. The clustering of sediments in PQR segment in C-M plotting indicates the estuarine characteristics of the region.

Keywords Depositional environment · Linear discriminate analysis B-variate plotting · Hydrodynamic processes · C-M plotting

3.1 Introduction

During recent years, mounting attentions have been given to renovate Palaeo-flow and sediment dispersal pattern based on the regular measurements of both directional as well as scalar quantities (Potter and Pettijhon 1977). It is a recognized and well accepted fact that the grain size of clastic sediments relates to the physical characteristics of the *environment of sediment deposition*, especially the dynamic forces working during deposition (Wentworth 1922; Krumbein and Pettijhon 1938;

Visher 1969; Tucker and Vacher 1980; Mc Laren 1981; Forrest and Clark 1989; Sahu 1964; Beal and Shepard 1956; Bradley 1999; Shepard 1960). Large number of studies has analyzed the grain size properties which can indicate the sources and the hydrodynamic characteristics of marine sediments (Carranza-Edwards et al. 2005). Different grain size parameters like mean (M) and standard deviation (SD) indicate the energy conditions of the environment of the sediment deposition (Visher 1969; Sly et al. 1982) because the variation in the distribution of sediment grain size is mainly due to the variation of wave energy reaching the point of sediment deposition and the extent of turbulence which affect the environment. Mason and Folk (1958), Folk (1966), Friedman (1961, 1967) and Moiola and Weiser (1968) explained that coarser sediments are deposited *in high-energy environments* while the deposition of finer sediments is observed in *low energy conditions*. Friedman (1961) and Moiola and Weiser (1968) suggested that bi-variate plotting of skewness and mean (phi value) of sediment grains reflects the obvious and complete separation of dune sands, river sands and ocean beach sands. They mentioned that environment of the deposition of sediments can be differentiated by the bi-variate plotting of kurtosis and skewness of sediment particles against each other. Bi-variate plotting of sorting of sediment grains against skewness is very useful to signify the difference between river sands from beach sands (Friedman 1961). Duane (1964) and Mason and Folk (1958) explained that Kurtosis is one of the important textural parameters to distinguish various environments of deposition of sediments. Passega (1957) and Passega and Byramjee (1969) used the *C-M plotting* widely to evaluate the *hydrodynamic processes* working during the deposition of the sediments. Ramanathan et al. (2009) mentioned that on the C-M plotting, concentration of points in PQR segment reflects the estuarine characteristics of the sediments.

3.2 Materials and Methodology of the Study

Sieving technique is used for the textural classification of sediments and graphical method of Folk and Ward (1957) was followed and different size parameters (Mean, Sorting, Skewness and Kurtosis) were calculated. *Linear Discriminate Analysis (LDA)* is done using these parameters to explain the environment of deposition (Sahu 1964).

$$Y1 = -3.5688\,\text{Mean} + 3.7016\,(\text{Standard Deviation})^2 - 2.0766\,\text{Skewness} + 3.1135\,\text{Kurtosis} \tag{3.1}$$

$$Y2 = 15.6534\,\text{Mean} + 65.7091\,(\text{Standard Deviation})^2 + 18.1071\,\text{Skewness} + 18.5043\,\text{Kurtosis} \tag{3.2}$$

3.2 Materials and Methodology of the Study

$$Y3 = 0.2852 \text{ Mean} - 8.7604 \text{ (Standard Deviation)}^2 - 4.8932 \text{ Skewness} + 0.0482 \text{ Kurtosis} \quad (3.3)$$

$$Y4 = 4.5129 \text{ Mean} - 1.2837 \text{ (Standard Deviation)}^2 + 3.5904 \text{ Skewness} + 4.1038 \text{ Kurtosis} \quad (3.4)$$

where, Y1 ≥ −2.7411 indicates Beach condition and Y1 < −2.7411 indicate Aeolian condition. Y2 ≥ 65.36 indicates Shallow agitated condition and Y2 < 65.36 indicate Beach condition. Y3 ≥ 7.41 indicates Shallow marine environment and Y3 < −7.41 indicates Fluvial environment. Y4 ≥ 9.81 indicates Fluvial deposit while Y4 < 9.81 indicates the impact of Turbidity current.

As the technique of Y1 and Y2 are applied for discrimination between beach from aeolian sand deposits and beach from shallow agitated condition respectively, these are not applicable for the present situation. So, Y1 and Y2 are avoided.

Depositional environment was assessed through bi-variate plotting of Mean versus Standard deviation and Standard deviation versus Skewness, following Friedman (1967) and Moiola and Weiser (1968). Hydrodynamic processes working during the deposition of the sediments have been identified by *C-M plotting* after Passega (1957) and Passega and Byramjee (1969).

3.3 Linear Discriminate Analysis (LDA) of Sediments

Linear Discriminate Analysis is an important measure to identify the environment of sediment deposition. Environment of sediment deposition can be interpreted from grain size analysis of sediments (Beal and Shepard 1956; Bradley 1999; Shepard 1960). According to Sahu (1964), the fluctuations of energy and fluidity factors have obvious correlation with the different operating processes and the environment of sediment deposition.

3.3.1 Environment of Sediment Deposition at Kolaghat

Approximately, 40% of the sediment samples showed Y3 values falling in shallow marine environment and remaining 60% samples in fluvial environment (Table 3.1). During monsoon season most of the samples (80%) fall in fluvial environment. During dry season 55% of the sediment samples indicate shallow marine environment and remaining 45% samples indicate fluvial environment. The Y4 values show that, about 60% of the samples are deposited by fluvial action and 40% by turbidity action (Table 3.1). So, the influence of fluvial and marine environment is almost equal in pre-monsoon and post-monsoon seasons, but during monsoon season voluminous riverine discharge helps the riverine influence to dominate over the marine influence (Maity 2015).

Table 3.1 Result of linear discriminate analysis of sediments at Kolaghat

Sediment sample	Pre-monsoon season		Monsoon season		Post-monsoon season	
	Y3	Y4	Y3	Y4	Y3	Y4
1	−7.55 (fluvial)	9.95 (fluvial deposit)	−9.18 (fluvial)	9.56 (turbidity current)	−9.84 (fluvial)	8.75 (turbidity current)
2	−7.25 (shallow marine)	7.16 (turbidity current)	−10.63 (fluvial)	845 (turbidity current)	−8.45 (fluvial)	9.59 (turbidity current)
3	−7.27 (shallow marine)	10.45 (fluvial deposit)	−8.29 (fluvial)	7.89 (turbidity current)	−6.27 (shallow marine)	8.93 (fluvial deposit)
4	−8.44 (fluvial)	9.89 (fluvial deposit)	−7.95 (fluvial)	9.94 (fluvial deposit)	−7.11 (shallow marine)	8.77 (turbidity current)
5	−8.14 (fluvial)	11.34 (fluvial deposit)	−8.63 (fluvial)	10.74 (fluvial deposit)	−8.74 (fluvial)	10.39 (fluvial deposit)
6	−9.15 (fluvial)	10.85 (fluvial deposit)	−6.93 (shallow marine)	9.93 (fluvial deposit)	−8.69 (fluvial)	10.18 (fluvial deposit)
7	−6.34 (shallow marine)	9.99 (fluvial deposit)	−7.79 (fluvial)	9.45 (turbidity current)	−6.98 (shallow marine)	9.93 (fluvial deposit)
8	−7.15 (shallow marine)	9.64 (turbidity current)	−10.13 (fluvial)	11.87 (fluvial deposit)	−7.24 (shallow marine)	10.55 (fluvial deposit)
9	−6.55 (shallow marine)	7.56 (turbidity current)	−9.55 (fluvial)	8.43 (turbidity current)	−8.97 (fluvial)	11.25 (fluvial deposit)
10	−6.37 (shallow marine)	8.94 (turbidity current)	−7.28 (shallow marine)	7.62 (turbidity current)	−7.15 (shallow marine)	6.36 (turbidity current)

Source Field survey and laboratory experiment

3.3.2 Environment of Sediment Deposition at Soyadighi

Though the influence of fluvial and marine environment is almost equal taking all the seasons as a whole but seasonal variation of this impact is noticeable. During monsoon season most of the samples (70%) fall in fluvial environment, but in dry season, 60% of the sediment samples indicate the dominance of shallow marine environment (Table 3.2). During freshet, huge amount of terrestrial discharge enhances the dominance of riverine influence over the marine influence (Maity 2015). Approximately, 52% of the samples are deposited by fluvial action and 48% by turbidity action as indicated by the Y4 values (Table 3.2).

3.3 Linear Discriminate Analysis (LDA) of Sediments

Table 3.2 Result of linear discriminate analysis of sediments at Soyadighi

Sediment sample	Pre-monsoon season		Monsoon season		Post-monsoon season	
	Y3	Y4	Y3	Y4	Y3	Y4
11	−7.25 (shallow marine)	8.76 (turbidity current)	−9.18 (fluvial)	8.66 (turbidity current)	−9.84 (fluvial)	10.75 (fluvial deposit)
12	−7.95 (fluvial)	10.48 (fluvial deposit)	−10.63 (fluvial)	8.77 (turbidity current)	−8.45 (fluvial)	12.88 (fluvial deposit)
13	−7.27 (shallow marine)	10.12 (fluvial deposit)	−8.29 (fluvial)	10.22 (fluvial deposit)	−6.27 (shallow marine)	7.92 (turbidity current)
14	−8.94 (fluvial)	9.98 (fluvial deposit)	−7.95 (fluvial)	8.17 (turbidity current)	−9.19 (fluvial)	10.54 (fluvial deposit)
15	−814 (fluvial)	11.42 (fluvial deposit)	−6.33 (shallow marine)	7.64 (turbidity current)	−8.74 (fluvial)	9.91 (fluvial deposit)
16	−715 (shallow marine)	7.84 (turbidity current)	−7.79 (fluvial)	10.12 (fluvial deposit)	−8.69 (fluvial)	11.42 (fluvial deposit)
17	−6.34 (shallow marine)	10.45 (fluvial deposit)	−6.93 (shallow marine)	9.99 (fluvial deposit)	−6.88 (shallow marine)	7.89 (turbidity current)
18	−7.15 (shallow marine)	9.79 (turbidity current)	−10.13 (fluvial)	11.32 (fluvial deposit)	−7.24 (shallow marine)	5.67 (turbidity current)
19	−6.55 (shallow marine)	10.78 (fluvial deposit)	−9.75 (fluvial)	8.56 (turbidity current)	−6.97 (shallow marine)	10.31 (fluvial deposit)
20	−6.37 (shallow marine)	9.23 (turbidity current)	−7.28 (shallow marine)	9.23 (turbidity current)	−7.35 (shallow marine)	9.32 (turbidity current)

Source Field survey and laboratory experiment

3.3.3 Environment of Sediment Deposition at Anantapur

Taking all the seasons together, 51% of the sediment samples show the influence of shallow marine environment and remaining 49% samples are affected by fluvial environment (Table 3.3). In monsoon season 60% of the sediment samples fall in fluvial environment, but in dry season 60% of the samples are affected by shallow marine environment. During monsoon season voluminous riverine discharge dominates over the marine influence (Maity 2015). The deposition of sediment is almost equally affected by the fluvial action and turbidity action, as indicated by the Y4 values (Table 3.3).

Table 3.3 Result of linear discriminate analysis of sediments at Anantapur

Sediment sample	Pre-monsoon season		Monsoon season		Post-monsoon season	
	Y3	Y4	Y3	Y4	Y3	Y4
21	−7.26 (shallow marine)	12.65 (fluvial deposit)	−9.72 (fluvial)	5.22 (turbidity current)	−7.39 (shallow marine)	8.70 (turbidity current)
22	−6.33 (shallow marine)	10.54 (fluvial deposit)	−6.52 (shallow marine)	7.85 (turbidity current)	−5.57 (shallow marine)	11.91 (fluvial deposit)
23	−10.11 (fluvial)	8.43 (turbidity current)	−6.71 (shallow marine)	14.65 (fluvial deposit)	−13.17 (fluvial)	14.04 (fluvial deposit)
24	−7.15 (shallow marine)	10.14 (fluvial deposit)	−10.16 (fluvial)	12.83 (fluvial deposit)	−6.72 (shallow marine)	14.59 (fluvial deposit)
25	−8.99 (fluvial)	13.64 (fluvial deposit)	−9.84 (fluvial)	9.76 (turbidity current)	−12.64 (fluvial)	13.27 (fluvial deposit)
26	−7.38 (shallow marine)	7.55 (turbidity current)	−12.65 (fluvial)	14.18 (fluvial deposit)	−5.97 (shallow marine)	9.74 (turbidity current)
27	−9.92 (fluvial)	11.73 (fluvial deposit)	−11.09 (fluvial)	9.33 (turbidity current)	−10.95 (fluvial)	14.12 (fluvial deposit)
28	−7.18 (shallow marine)	7.09 (turbidity current)	−5.43 (shallow marine)	7.84 (turbidity current)	−9.16 (fluvial)	8.79 (turbidity current)
29	−11.54 (fluvial)	11.18 (fluvial deposit)	−4.12 (shallow marine)	6.92 (turbidity current)	−4.77 (shallow marine)	10.87 (fluvial deposit)
30	−7.52 (shallow marine)	6.94 (turbidity current)	−11.75 (fluvial)	5.95 (turbidity current)	−5.69 (shallow marine)	9.04 (turbidity current)

Source Field survey and laboratory experiment

3.3.4 Environment of Sediment Deposition at Pyratungi

Y3 values indicate that 53% of the samples are affected by shallow marine environment and 47% samples are affected by fluvial environment. The influence of fluvial and marine processes varies seasonally. In monsoon season, 60% of the sediment samples fall under fluvial environment, but in dry season, 60% of the sediment samples indicate the influence of shallow marine environment (Table 3.4). During rainy season, the riverine influence is more than that of marine (Maity 2015). Approximately, 54% of the samples are deposited under weak fluvial action, mainly in dry season and 46% by turbidity action as indicated by the Y4 values (Table 3.4).

3.3 Linear Discriminate Analysis (LDA) of Sediments

Table 3.4 Result of linear discriminate analysis of sediments at Pyratungi

Sediment sample	Pre-monsoon season		Monsoon season		Post-monsoon season	
	Y3	Y4	Y3	Y4	Y3	Y4
31	−10.47 (fluvial)	11.73 (fluvial deposit)	−12.84 (fluvial)	7.84 (turbidity current)	−7.23 (shallow marine)	12.83 (fluvial deposit)
32	−4.92 (shallow marine)	9.94 (fluvial deposit)	−11.36 (fluvial)	9.95 (fluvial deposit)	−10.55 (fluvial)	11.89 (fluvial deposit)
33	−1183 (fluvial)	10.56 (fluvial deposit)	−6.83 (shallow marine)	8.77 (turbidity current)	−6.82 (shallow marine)	11.09 (fluvial deposit)
34	−6.73 (shallow marine)	11.43 (fluvial deposit)	−6.93 (shallow marine)	7.94 (turbidity current)	−14.26 (fluvial)	9.48 (turbidity current)
35	−7.25 (shallow marine)	10.66 (fluvial deposit)	−10.48 (fluvial)	12.09 (fluvial deposit)	−6.93 (shallow marine)	11.07 (fluvial deposit)
36	−6.56 (shallow marine)	5.26 (turbidity current)	−9.85 (fluvial)	8.82 (turbidity current)	−12.93 (fluvial)	8.73 (turbidity current)
37	−14.35 (fluvial)	14.22 (fluvial deposit)	−12.47 (fluvial)	6.36 (turbidity current)	−11.25 (fluvial)	8.29 (turbidity current)
38	−7.19 (shallow marine)	12.94 (fluvial deposit)	−6.83 (shallow marine)	10.94 (fluvial deposit)	−6.98 (shallow marine)	12.99 (fluvial deposit)
39	−5.93 (shallow marine)	8.46 (turbidity current)	−7.15 (shallow marine)	8.35 (turbidity current)	−5.38 (shallow marine)	13.29 (fluvial deposit)
40	−10.99 (fluvial)	7.59 (turbidity current)	−13.28 (fluvial)	7.83 (turbidity current)	−7.25 (shallow marine)	7.94 (turbidity current)

Source Field survey and laboratory experiment

3.3.5 Environment of Sediment Deposition at Dhanipur

Taking all the sediment samples together, nearly 61% of the samples indicate the impact of shallow marine environment and remaining 39% samples are affected by fluvial environment (Table 3.5), which reflect the downstream increase of marine impact. During dry season most of the sediment samples (65%) are affected by shallow marine environment, while in monsoon season 60% samples indicate the impact of fluvial processes. During monsoon season voluminous riverine discharge enhances the riverine influence over the marine influence (Maity 2015). The Y4 values show that, about 64% of the samples are deposited under weak fluvial action, mainly in dry season and remaining 36% by turbidity action (Table 3.5).

Table 3.5 Result of linear discriminate analysis of sediments at Dhanipur

Sediment sample	Pre-monsoon season		Monsoon season		Post-monsoon season	
	Y3	Y4	Y3	Y4	Y3	Y4
41	−6.93 (shallow marine)	12.76 (fluvial deposit)	−6.84 (shallow marine)	7.62 (turbidity current)	−7.39 (shallow marine)	9.74 (turbidity current)
42	−7.35 (shallow marine)	9.85 (fluvial deposit)	−4.38 (shallow marine)	11.91 (fluvial deposit)	−5.87 (shallow marine)	10.59 (fluvial deposit)
43	−12.28 (fluvial)	11.79 (fluvial deposit)	−8.25 (fluvial)	11.81 (fluvial deposit)	−10.69 (fluvial)	12.74 (fluvial deposit)
44	−7.39 (shallow marine)	11.55 (fluvial deposit)	−11.77 (fluvial)	14.56 (fluvial deposit)	−11.84 (fluvial)	11.27 (fluvial deposit)
45	−5.55 (shallow marine)	7.83 (turbidity current)	−4.92 (shallow marine)	8.94 (turbidity current)	−13.36 (fluvial)	14.52 (fluvial deposit)
46	−12.65 (fluvial)	12.84 (fluvial deposit)	−13.83 (fluvial)	6.78 (turbidity current)	−6.85 (shallow marine)	7.84 (turbidity current)
47	−6.76 (shallow marine)	5.63 (turbidity current)	−6.73 (shallow marine)	7.84 (turbidity current)	−6.93 (shallow marine)	11.23 (fluvial deposit)
48	−6.23 (shallow marine)	8.77 (turbidity current)	−7.98 (fluvial)	5.82 (turbidity current)	−14.48 (fluvial)	9.94 (fluvial deposit)
49	−14.93 (fluvial)	12.84 (fluvial deposit)	−11.56 (fluvial)	12.59 (fluvial deposit)	−4.85 (shallow marine)	12.93 (fluvial deposit)
50	−7.35 (shallow marine)	11.98 (fluvial deposit)	−14.35 (fluvial)	8.95 (turbidity current)	−4.44 (shallow marine)	8.48 (turbidity current)

Source Field survey and laboratory experiment

3.3.6 Environment of Sediment Deposition at Geonkhali

Nearly, 60% of the sediment samples show the influence of shallow marine environment and remaining 40% samples are affected by fluvial environment (Table 3.6), which reflect the downstream increase of marine impact. During monsoon season the impact of fluvial and marine processes is almost equal, but in dry season 65% of the samples are affected by shallow marine environment. During monsoon season voluminous riverine discharge causes the enhancement of the riverine influence over the marine influence (Maity 2015). The deposition of sediment is mostly affected by fluvial action than by turbidity action, as indicated by the Y4 values (Table 3.6).

Table 3.6 Result of linear discriminate analysis of sediments at Geonkhali

Sediment Sample	Pre-monsoon season		Monsoon season		Post-monsoon season	
	Y3	Y4	Y3	Y4	Y3	Y4
51	−10.23 (fluvial)	10.35 (fluvial deposit)	−13.84 (fluvial)	12.43 (fluvial deposit)	−10.32 (fluvial)	11.24 (fluvial deposit)
52	−9.51 (fluvial)	12.83 (fluvial deposit)	−7.12 (shallow marine)	10.72 (fluvial deposit)	−7.12 (shallow marine)	14.31 (fluvial deposit)
53	−6.65 (shallow marine)	12.54 (fluvial deposit)	−14.55 (fluvial)	8.23 (turbidity current)	−6.77 (shallow marine)	10.54 (fluvial deposit)
54	−12.34 (fluvial)	10.63 (fluvial deposit)	−6.99 (shallow marine)	11.68 (fluvial deposit)	−6.59 (shallow marine)	7.58 (turbidity current)
55	−7.13 (shallow marine)	11.54 (fluvial deposit)	−5.64 (shallow marine)	9.36 (turbidity current)	−5.86 (shallow marine)	12.56 (fluvial deposit)
56	−6.55 (shallow marine)	9.36 (turbidity current)	−6.27 (shallow marine)	8.57 (turbidity current)	−10.54 (fluvial)	10.57 (fluvial deposit)
57	−6.73 (shallow marine)	7.36 (turbidity current)	−13.41 (fluvial)	8.45 (turbidity current)	−12.39 (fluvial)	11.57 (fluvial deposit)
58	−5.38 (shallow marine)	6.39 (turbidity current)	−11.53 (fluvial)	9.43 (turbidity current)	−5.62 (shallow marine)	9.25 (turbidity current)
59	−7.35 (shallow marine)	7.83 (turbidity current)	−15.97 (fluvial)	8.44 (turbidity current)	−6.73 (shallow marine)	12.67 (fluvial deposit)
60	−15.43 (fluvial)	14.61 (fluvial deposit)	−7.36 (shallow marine)	8.37 (turbidity current)	−5.11 (shallow marine)	7.41 (turbidity current)

Source Field survey and laboratory experiment

3.4 Influences of Fluvial and Marine Processes

Being a part of *Hoogly estuary*, the *mechanism of sedimentation* at lower reach of the Rupnarayan River is the result of the interaction of riverine and marine processes. The degree of interaction and dominance of these two processes depend on several factors. In riverine flow, the volume and discharge of water is important; while in marine flow the tidal range, tidal asymmetry and tidal prism play the important role to control this interaction. Seasonal fluctuation of stream velocity and energy, channel diversion and flow separation, generation of turbulence, the mixing of fresh river water and marine saline water and the loss of velocity of river water etc. have a combined effect on the processes and mechanisms of sedimentation in the area under study (Perkins 1974).

3.4.1 Spatial and Seasonal Variation of Influences of Fluvial and Marine Processes

The influences of fluvial and marine processes undergo changes seasonally at different places along the lower reach of the Rupnarayan River. During pre-monsoon and post-monsoon seasons the riverine discharge is insufficient due to paucity of rainfall. As a result, the *marine environment* becomes dominant over the *fluvial environment* (Fig. 3.1a, c). But during monsoon season, occurrence of huge rainfall increases the riverine discharge and reduces the influence of marine processes (Fig. 3.1b). The influence of fluvial and marine processes also varies spatially. In all the seasons the influence of fluvial processes has decreased and marine influence has increased gradually towards downstream (Fig. 3.1a–c). The seasonal and spatial variation of fluvial and marine influences leads to the fluctuation of stream energy and affect the mechanism of sedimentation in the study area (Maity 2015).

3.5 Bi-variate Plot of Mean and Standard Deviation

The relationship between different sediment size-parameters is very important to understand various aspects of environment of sediment deposition, because the grain size parameters of the sediments are extremely environmentally sensitive (Folk and Ward 1957; Passega 1957; Friedman 1961, 1967; Moiola and Weiser 1968; Visher 1969). The plot between mean vs standard deviation is considered as an effective tool to differentiate between beach and river sands (Moiola and Weiser 1968; Friedman 1967). From the bi-variate plots of mean versus standard deviation, it is seen that in pre-monsoon and post-monsoon more than 60% of the sediment samples (Fig. 3.2a, c) fall in beach environment, indicating the dominance of marine influence over the riverine influence. But during monsoon season voluminous upland discharge enhance the riverine influence over the tidal influence. More than 65% of the sediment samples (Fig. 3.2b) fall under riverine environment in this season, indicating the dominance of riverine influence over the marine influence.

3.5 Bi-variate Plot of Mean and Standard Deviation

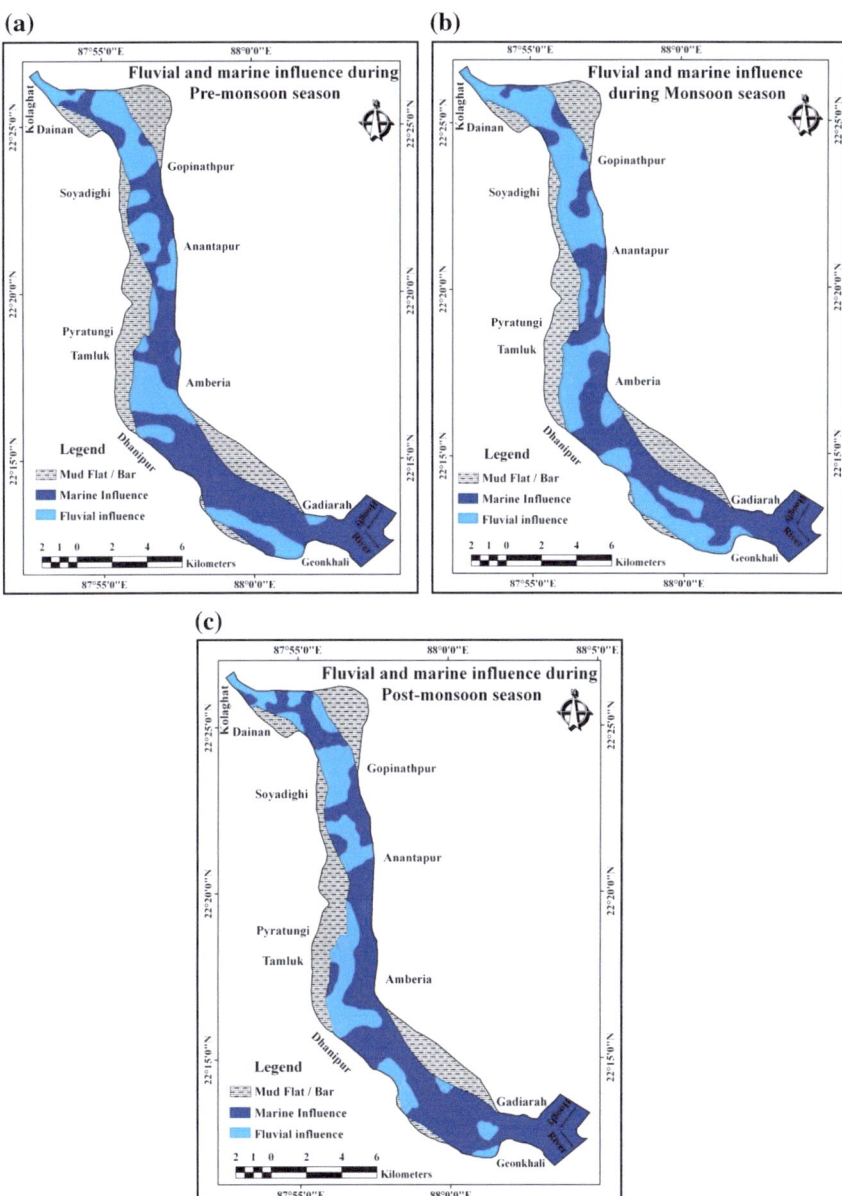

Fig. 3.1 **a** Fluvial and marine influence in pre-monsoon season. **b** Fluvial and marine influence in monsoon season. **c** Fluvial and marine influence in post-monsoon season. *Source* Field survey and laboratory experiment

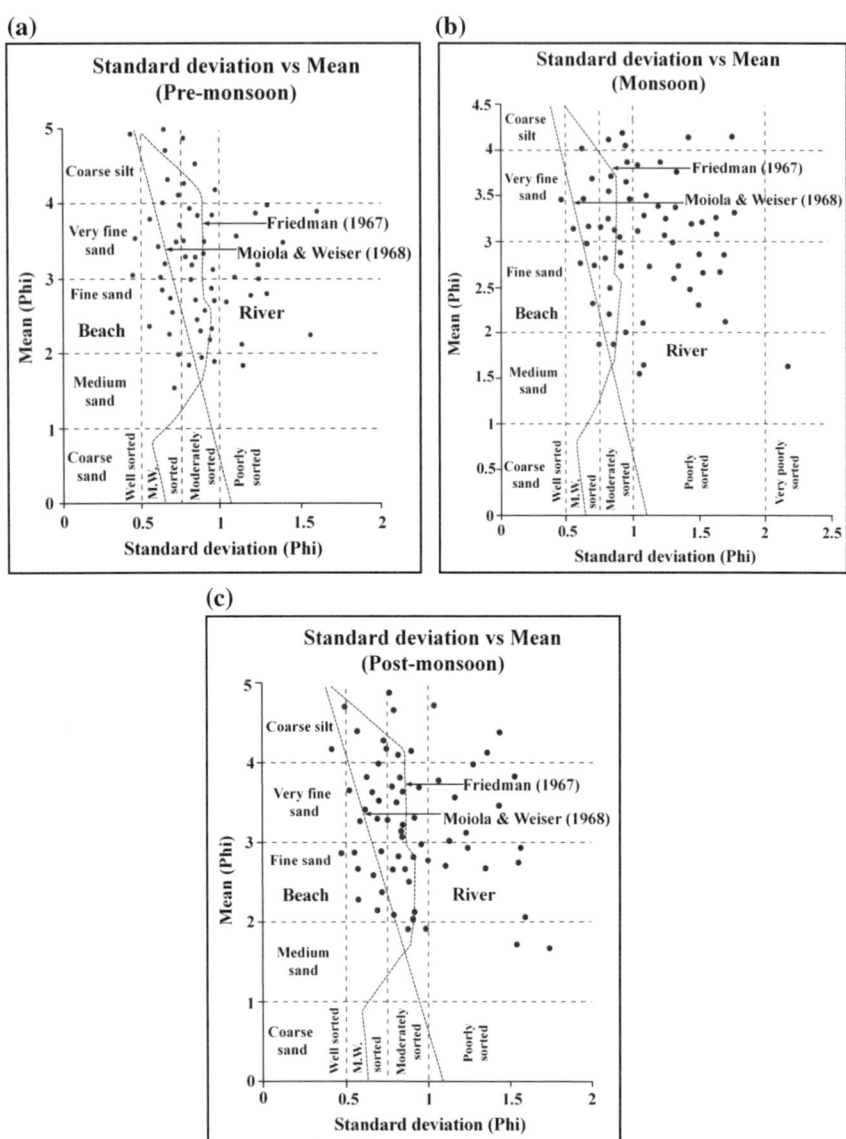

Fig. 3.2 a Bi-variate plot of mean versus standard deviation in pre-monsoon season. b Bi-variate plot of mean versus standard deviation in monsoon season. c Bi-variate plot of mean versus standard deviation in post-monsoon season. *Source* Field survey and laboratory experiment

3.6 Bi-variate Plot of Skewness and Standard Deviation

Friedman (1967), Moiola and Weiser (1968) considered that bi-variate plotting of standard deviation (along horizontal axis) and skewness (along vertical axis) is very useful and significant for the differentiation between beach environment and riverine environment. Plotting of standard deviation vs skewness of sediments indicates that, more than 62% of the sediment samples during pre-monsoon and post-monsoon seasons fall in beach environment and remaining sediments are deposited in riverine environment (Fig. 3.3a, c). This represents the dominance of marine influence over the riverine influence in dry season. But during freshet, the environment of sediment deposition is different from that in dry season. In monsoon season, huge volume of upland discharge enhances the riverine influence over the marine influence. More than 65% of the sediment samples (Fig. 3.3b) fall under riverine environment in this season. So, during monsoon season the plot shows the concentration of points in the river domain as proposed by Friedman (1967).

3.7 Deflection of Marine and Riverine Flow by Coriolis Force

The influence of the earth's rotation (Coriolis force) tends to swing the salt-water flow and the fresh-water flow towards their right (in the Northern Hemisphere) (Pethick 1984). River discharge (ebb tide) and flood tide move in opposite direction to each other. Marine salt water flows along the left bank of the estuary (facing towards the sea) but the riverine fresh water hugs towards the right hand bank (Fig. 3.4). Mixing between salt water and fresh water takes place laterally which makes the tidal current more powerful. Flood tide carries marine sediments into the estuary, which are deposited on the left bank while the weaker river flow deposits upland sediment on the right bank (Pethick 1984). In the study area the rate of sedimentation is high towards the right bank of the river (Fig. 3.5), mainly during pre-monsoon and post-monsoon seasons. It is due to the fact that in pre and post-monsoon the riverine flow becomes weak (lack of upland discharge), causing the reduction of sediment transporting capacity and thus rapid sedimentation towards the right bank becomes possible. In contrast, on the left bank strong tidal current leads to easy transportation of sediments and so, sedimentation rate is reduced. During monsoon season both riverine and marine flow are strong, which enhances the energy and sediment transporting capacity and the sedimentation rate is reduced to a great extent.

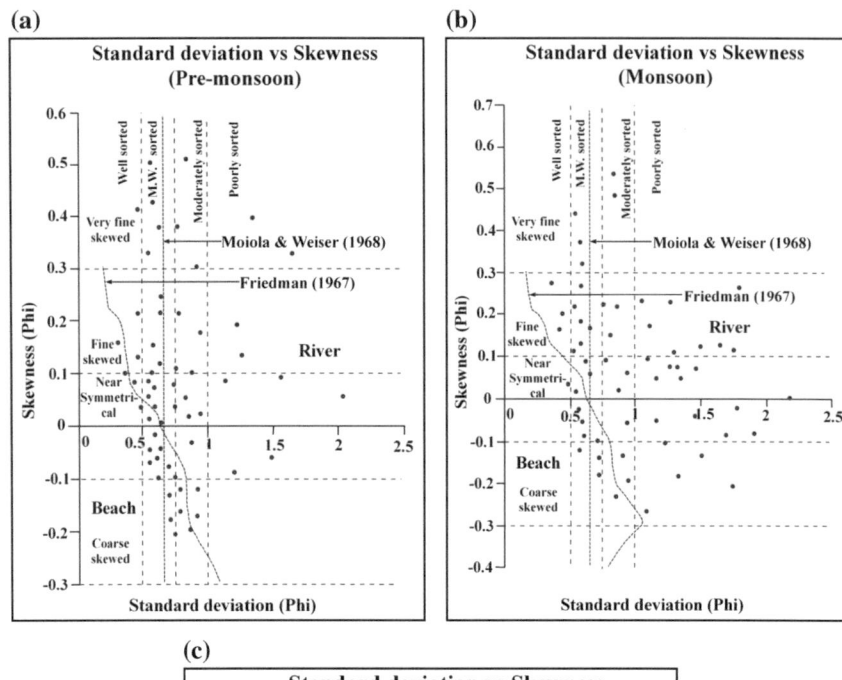
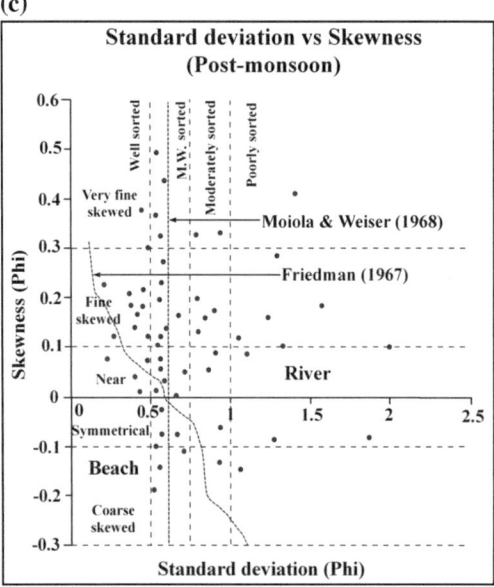

Fig. 3.3 a Bi-variate plot of standard deviation versus skewness in pre-monsoon season. b Bi-variate plot of standard deviation versus skewness in monsoon season. c Bi-variate plot of standard deviation versus skewness in post-monsoon season. *Source* Field survey and laboratory experiment

Fig. 3.4 Deflection of marine and riverine flow by Coriolis force (Pethick 1984)

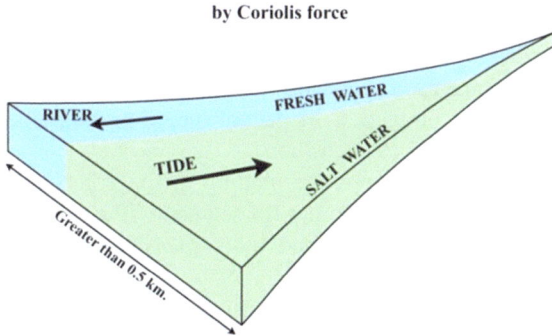

Fig. 3.5 More sedimentation towards right bank due to rightward deflection of weaker riverine flow by Coriolis force

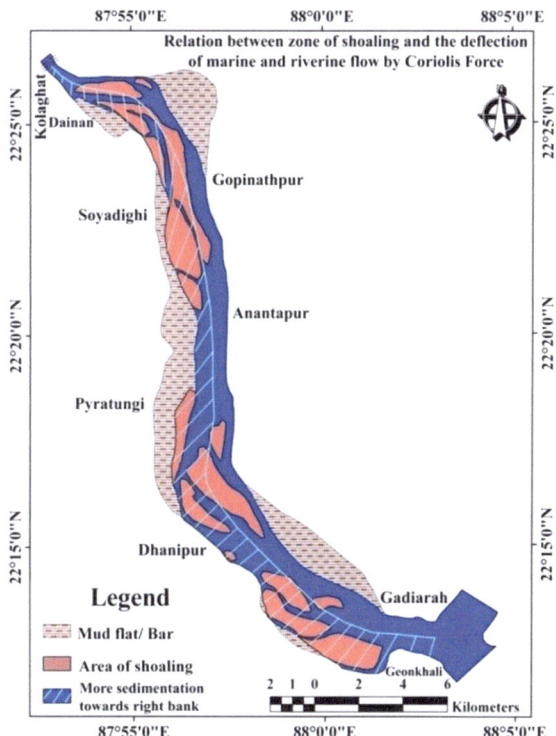

3.8 C-M Plotting to Identify Hydrodynamic Forces Working During Deposition

The *C-M pattern* is an important platform to understand the modes transport of sediments before their deposition (Passega 1957; Passega and Byramjee 1969). It is a relationship of coarser one percentile value in micron (C) and median value in micron (M) of sediments on a log-probability scale. The area of a complete C-M pattern can be divided into several segments such as NO, OP, PQ, QR and RS indicating different modes of transport of sediments and various sedimentary environments. The location of the plotted points for a sediment deposit within the area of a complete C-M pattern indicates the probable conditions of transport of sediments before deposition.

In the study area, the plotting of the points represents that the sediments are populated in PQ, QR and RS segments (Fig. 3.6) which indicates that most of the sediments are transported by *suspension with rolling, graded suspension and uniform suspension* (Passega 1957; Passega and Byramjee 1969). Nearly, 89% of the sediments are transported jointly by graded suspension and suspension with rolling. Uniform suspension transport of sediment is mostly (87.5%) observed in pre-monsoon and post-monsoon season which indicates the dominants of finer sediments in dry season. The median value of the sediment samples ranges between 15 and 300 μ (Fig. 3.6). The value of first percentile of 97% of the sediment samples ranges between 200 and 1000 μ (Fig. 3.6) which infer *moderate to lesser violent hydrodynamic condition* leading to deposition (Passega 1957; Passega and Byramjee 1969). Only 3% of the sediments (mostly in monsoon season) have one percentile value greater than 1000, indicating moderately high violent hydrodynamic condition of sediment deposition. This group of sediment reflects suspension with rolling, graded suspension as well as uniform suspension mode of transportation history, indicating the complexity in the hydrodynamic processes operating in this system.

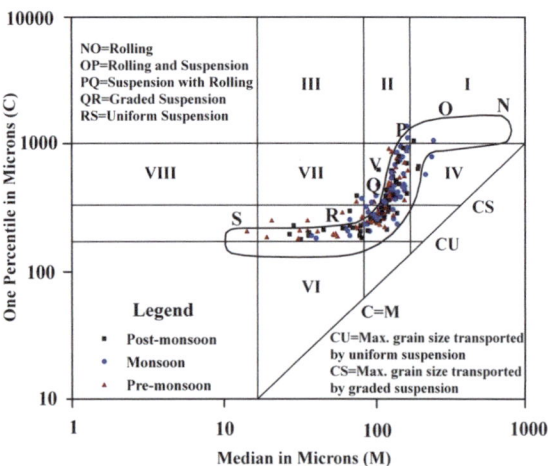

Fig. 3.6 C-M pattern of sediments. *Source* Field survey

References

Beal M, Shepard F (1956) A use of roundness to determine depositional environments. J Sediment Petrol 26:9–60

Bradley WC (1999) Submarine erosion and wave cut platforms. Bull Geol Soc Am 69:967–974

Carranza-Edwards A, RosalesHoz L, Urrutia-Fucugauchi J et al (2005) Geochemical distribution pattern of sediments in an active continental shelf in Southern Mexico. Cont Shelf Res 25: 521–537

Duane DB (1964) Significance of skewness in recent sediments, western Pamlico Sound. North Carol J Sediment Petrol 34(4):864–874

Folk RL (1966) A review of grain-size parameters. Sedimentology 6:73–93

Folk RL, Ward MC (1957) Brazos River bar (Texas): a study in the significance of grain size parameters. J Sediment Petrol 27(1):3–27

Forrest J, Clark NR (1989) Characterizing grain size distribution: evaluation of a new approach using multivariate extension of entropy analysis. Sedimentology 36:711–722

Friedman GM (1961) Distinction between dune, beach and river sands from their textural characteristics. J Sediment Petrol 31(4):514–529

Friedman GM (1967) Dynamic processes and statistical parameters compared for size frequency distribution of beach river sands. J Sediment Petrol 37(2):327–354

Krumbein WC, Pettijhon EJ (1938) Manual of sedimentary petrology. Applton-Century Crofts, New York

Maity SK (2015) Cognition of interworking of processes leading to sedimentation at lower reach of the Rupnarayan River, West Bengal, India. Dissertation, Vidyasagar University, West Bengal, India

Mason CC, Folk RL (1958) Differentiation of beach, dune and aeolian flat environments by size analysis, Mustang Island, Texas. J Sediment Petrol 28:211–226

Mc Laren P (1981) An interpretation of trends in grain size measures. J Sediment Petrol 51: 611–624

Moiola RJ, Weiser D (1968) Textural parameters: an evaluation. J Sediment Petrol 38(1):45–53

Passega R (1957) Grain size representation by CM patterns as a geologic tool. J Sediment Petrol 34:830–857

Passega R, Byramjee R (1969) Grain size image of clastic deposits. Sedimentology 13:180–190

Perkins FJ (1974) The biology of estuaries and coastal waters. Academic Press, London

Pethick JS (1984) An introduction to coastal geomorphology. Arnold, London

Potter PE, Pettijhon EJ (1977) Palaeocurrent and basin analysis. Springer

Ramanathan AL et al (2009) Textural characteristics of the surface sediments of a tropical mangrove sundarban ecosystem India. Indian J Mar Sci 38(4):397–403

Sahu BK (1964) Depositional mechanism from the size analysis of clastic sediments. J Sediment Petrol 34(1):73–83

Shepard FP (1960) Distinguishing between beach and dune sands. J Sediment Petrol 31:196–214

Sly PG, Thomas RL, Pelletier BR (1982) Comparison of sediment energy-texture relationships in marine and lacustrine environments. Hydrobiologia 91:71–84

Tucker RW, Vacher HL (1980) Effectiveness of discriminating beach, dune and river sands by moments and the cumulative weight percentages. J Sediment Petrol 50:165–172

Visher GS (1969) Grain size distributions and depositional processes. J Sediment Petrol 39 (3):1074–1106

Wentworth CK (1922) A scale of grade and class terms for sediments. J Geol 30:377–392

Chapter 4
Identification of the Sediment Sources Using X-Ray Diffraction (XRD) Technique

Abstract *X-ray diffraction (XRD)* technique is used to understand the *sources of sediments* through identification of *mineral composition* of sediments in the lower reach of the Rupnarayan River to explain the causes and mechanisms of sedimentation. A total of 21 sediment samples (13 samples from river bed and 8 samples from river banks) have been collected for knowing the sediment mineralogy. Sediment samples are washed by boiled distilled water, dried, disintegrated and scanned at an interval of 7°–45°2θ in XPERT-PRO diffractometer. *Diffractograms* produced by XRD study indicates that the entire lower reach shows the dominance of the minerals such as quartz, chlorite, illite, anatase, goethite, oligoclase, chloritoid, corundum, sillimanite, which have their origin in the upper and middle catchment area with small contribution from the lower catchment and river banks. Statistical experiment indicates that excluding tourmaline and anatase, all the minerals show steady trend in concentration in sediments. *Principal Component Analysis* (*PCA*) indicates that five Eigen values contribute for about 83.154% of the total variation of the distribution of minerals. The minerals discharged from the upper catchment are captured in the estuary and again redistributed towards upstream by stronger flood tide. This leads to a un-conspicuous and hapazard distribution of minerals in the area under study.

Keywords Sedimentation · Mineral composition · X-ray diffraction Sediment sources · PCA

4.1 Introduction

Most of the researchers and scientists have used *X-Ray Diffraction (XRD)*, an important instrumental analytical technique to identify the crystalline materials for more than a century (Tankersley and Balantyne 2010). This technique can be used to identify crystalline "fingerprints" by comparing the d-spacing (i.e., the distance between adjacent planes of atoms) of unknown samples with standard reference patterns and measurements (Chen 1977; Tankersley and Balantyne 2010).

Generally, the archaeologists have used this technique for the identification of the various clay minerals within different ancient pottery (Chase 1971; Matson 1971; Shepard 1971; Tankersley and Meinhart 1982; Tankersley and Balantyne 2010), efflorescent salts on biological artifacts (FitzHugh and Gettens 1971; Tankersley et al. 1985; Tankersley and Balantyne 2010), and the composition of the minerals of artifacts (Tankersley 1995; Tankersley et al. 1990, 1995; Tankersley and Balantyne 2010). In recent times, XRD has been used to reconstruct the past climates from fine-grained Late Holocene ponded sediments (Tankersley and Balantyne 2010). *Mineral composition* of sediments supplemented by the textural characteristics bear the nature of the source region and the depository environment (Rittenhouse 1943; Friedman 1961; Blatt et al. 1980). Different minerals are mostly deposited depending on the variations of their size, shape and density and thus a single size sediment fraction is unable to represent the overall mineralogical composition in the sediments (Tankersley and Balantyne 2010). Peak intensity ratios of different minerals in deposited sediment samples are good indicative of diverse clay source regions and hydro-meteorological transportation processes in the catchment areas (Trachsel et al. 2008; Tankersley and Balantyne 2010; Zong et al. 2015). The quantitative X-ray powder diffraction analysis of different clay minerals of large number of samples is based on the principles of the use of the standard powder samples and smear-oriented mounting techniques (Gibbs 1967). Both the techniques are based on the use of internal standard method. In this way the advantages of internal standard method (given by Klug and Alexander 1954) and highly sensitive smear-oriented technique for clay minerals were combined together (Gibbs 1965, 1967).

4.2 Significance of the Understanding of Sources of Sediments in the Study Area

Reduction of river bed sedimentation rate and management of associated problems are getting increasing attention in recent times from an economic, social and environmental perspective. In case of a river without tidal influence, sediments are mainly supplied from the upper catchment area and the local source. But in tidal river, the sediments which are being deposited supplied from two sources, i.e., from the catchment area of the river and from the beach and marine sources. The amount and volume of sediment supply from the river catchment area depends on the climatic character, seasonal distribution of rainfall, geological and geomorphological characteristics, topography and soil type, vegetation coverage and human interference on river basin etc. On the other hand the marine sediment supply depends on the degree of dominance of marine influence (tidal prism, tidal range and tidal asymmetry etc.) over the riverine influence. So, identification of the sources of sediments is the key to understand the *causes, mechanisms and dimension of sedimentation* and becomes helpful to reduce the rate of sedimentation and the management of the associated problems in the area of interest and in any of the river in the world. In the study area, no such studies have been done earlier by the researchers or the concerned authorities working in this purpose. This offers the opportunity to understand the *sources of sediments* through identification of *mineral composition* of sediments in the concerned area to have an insight into the causes and mechanisms of sedimentation and formulating strategies for proper management of the associated problems.

4.3 Mineralogy of the Catchment Area of the Rupnarayan River

Most of the region of the upper and middle catchment area of the Rupnarayan River is covered by *granite and gneiss* along with shale and sandstone of lower Gondwana system, micaschist, phyllite, quartzite, sandstone of upper Gondwana system, amphibolites, hornblende, grit and conglomerate (O'Malley 1995; Mukhopadhyay and Dasgupta 2010) (see Sect. 1.2.1 and Fig. 1.3 in Maity and Maiti 2018). All together, the upper and middle catchment area of the Rupnarayan River is composed of the minerals including *quartz*, goethite, illite, chlorite, chloritoid, mirabilite, chromite, dolomite, anatase, sillimanite, corundum, oligoclase, feldspar, mica, hornblende, augite, olivin, ilmenite, garnet, staurolite, tourmaline, calcite, gypsum, magnetite, apatite, plagioclase, amphibole, actinolite, zoisite and pyroxene (Maity and Maiti 2016).

The lower portion of the catchment area (mostly the eastern and southern portion) is mainly formed of *newer alluvium* but laterite, pleistocene sediments and older alluvium are also found here (O'Malley 1995; Mukhopadhyay and Dasgupta 2010) (Fig. 1.3 in Maity and Maiti 2018). The mineral compositions of the lateritic and alluvial soils of the lower part of the catchment of the Rupnarayan River have been studied by Ghosh and Datta (1974) using XRD technique. The study reveals that kaolinite (51–56%) and illite (17–22%) are the dominant minerals in the lateritic region. Chlorite (5–6%), smectite (4–6%), goethite (6–8%) and mixed layer minerals (8–10%) are also identified in little proportion in this soil. Smectite (29–31%) and illite (24–26%) minerals are the main constituents of older and newer alluvial soil. Quartz (8–10%), kaolinite (10–13%) and chlorite (10–14%) minerals are also identified in this soil. Goethite (3–6%), feldspar (3–4%) and mixed layer minerals (3–7%) are present in very little proportion. Table 4.1 shows the percentage amount of different minerals identified by Ghosh and Datta (1974) in the lower catchment area.

Table 4.1 Mineralogy of the lower catchment area (Ghosh and Datta 1974)

Soil sample	% amount of minerals							
	Kaolinite	Chlorite	Smectite	Illite	Quartz	Goethite	Feldspar	Mixed layer mineral
Sample-1 (laterite soil)	56	5	4	17	0	8	0	10
Sample-2 (laterite soil)	51	6	6	22	0	6	0	8
Sample-3 (alluvial soil)	13	11	29	25	9	3	3	7
Sample-4 (alluvial soil)	10	14	31	26	8	5	3	3
Sample-5 (alluvial soil)	12	10	29	24	10	6	4	5

4.4 Materials and Methodology of the Study

4.4.1 Treatments of Collected Sediment Samples

A total of *13 sediment samples* were collected from the river bed for mineralogical analysis: S1—from the upper reach (Bandar), S2—from the middle reach (Jasar) and remaining eleven (11) sediment samples from the lower reach [S3—from Kolaghat, S4—from Kantapukur, S5—from Gopinathpur, S6—from Soyadighi, S7—from Telipara, S8—from Anantapur, S9—from Pyratungi, S10—from Amberia, S11—from Dhanipur, S12—from Bholsara and S13—from Geonkhali] (Figs. 4.1 and 4.2a). Sediment samples were collected mostly from the area of shoaling in the lower reach.

Grain size analysis of the sediments reveal that samples S1, S2, S3 and S6 are of fine sand category, samples S4, S5, S7, S8, S9, S10, S12 and S13 are of very fine sand category and sample S11 is coarse silt in nature (Table 4.2). The proportion of sand, silt and clay ranges between 51 to 80%, 16 to 32% and 4 to 17% respectively in the sediment samples. Samples S10, S11 and S12 are of muddy sand in nature and remaining other samples are of silty sand type (Table 4.2).

In addition to this, other *8 sediment samples* were collected from the two banks of the lower reach of the Rupnarayan River to identify the mineral composition of the sediments in the river bank. Samples 1, 2, 3 and 4 have been taken from the left bank whereas samples 5, 6, 7 and 8 have been taken from the right bank of the river (Fig. 4.1). Sample 1, 2, 5 and 8 are of muddy sand, sample 6 is sandy clay and

4.4 Materials and Methodology of the Study

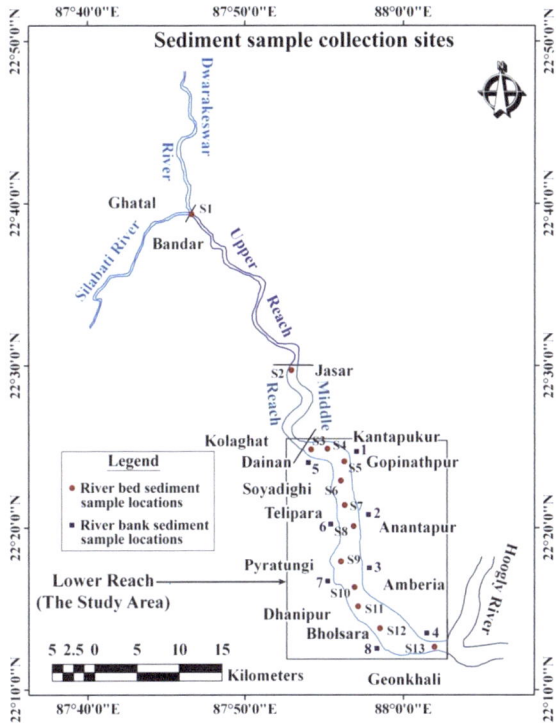

Fig. 4.1 Location of sediment samples collected for XRD analysis (Maity and Maiti 2016)

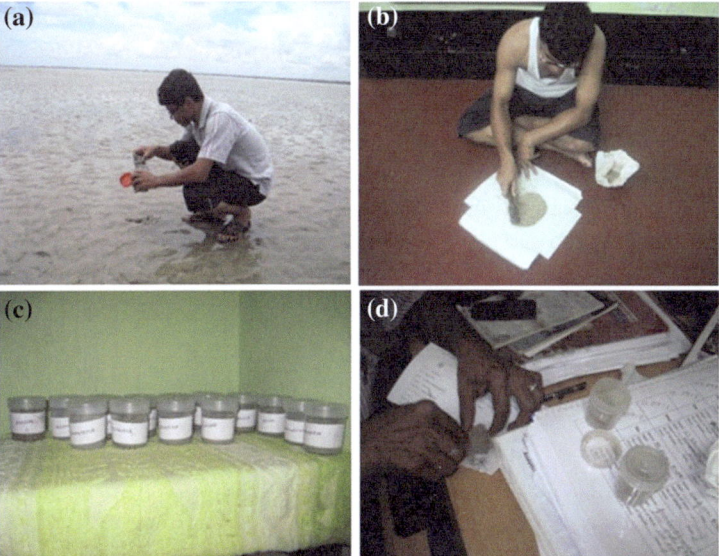

Fig. 4.2 Sediment sample collection (**a**), preparation of sediment samples (**b**), prepared sediment samples (**c**) and sample pressed in sample holder (**d**)

Table 4.2 Grain size characteristics of bed sediments (Maity and Maiti 2016)

Location	Mean grain size (mm)	Proportion of different particles (%)			Sediment type
		Sand	Silt	Clay	
Bandar (S1)	0.244	76	18	6	Silty sand
Jasar (S2)	0.175	80	16	4	Silty sand
Kolaghat (S3)	0.127	68	24	8	Silty sand
Kantapukur (S4)	0.116	67	20	13	Silty sand
Gopinathpur (S5)	0.104	70	21	9	Silty sand
Soyadighi (S6)	0.130	72	23	5	Silty sand
Telipara (S7)	0.109	65	23	12	Silty sand
Anantapur (S8)	0.085	61	26	13	Silty sand
Pyratungi (S9)	0.073	55	32	13	Silty sand
Amberia (S10)	0.071	53	32	15	Muddy sand
Dhanipur (S11)	0.061	51	32	17	Muddy sand
Bholsara (S12)	0.065	52	31	16	Muddy sand
Geonkhali (S13)	0.075	56	31	13	Silty sand

Source Field survey and laboratory experiment

sample 3, 4 and 7 are of clayey sand type. Amount of sand, silt and clay ranges between 40 to 66%, 12 to 36% and 22 to 55% respectively in sediment samples (Maity and Maiti 2016).

To reduce the seawater salt content below the detection level of X-ray diffraction and to remove the organic residues the collected sediments were washed with hot and distilled water. The washed sediment samples were dried and disintegrated manually with a mortar and pestle (Fig. 4.2b). Coarse grained sediments are breakdown up to silt size which allows the grain sizes of the crystallites for X-ray diffraction. Trihexylamine acetate was then mixed with the sediment samples to expand the smectite minerals (modified from Rex and Bauer 1965). The samples were allowed to dry for several days to attain equilibrium with normal humidity conditions. Trihexylamine acetate remains in the clay minerals for several weeks as it has a low vapor pressure (Fig. 4.2c). The equilibrated powdered sediment samples were then pressed into sample holders (Fig. 4.2d) by a custom-built sample press (Rex and Chown 1960) and analyzed by XRD technique using an XPERT-PRO diffractometer (Fig. 4.3a) at Central Research Facility, IIT, Kharagpur.

4.4.2 XRD Analysis of Sediment Samples

Mineralogical composition of the all sediment samples was determined by *diffractogram* (Figs. 4.3b and 4.4) generated from XRD study. The instrumental

4.4 Materials and Methodology of the Study

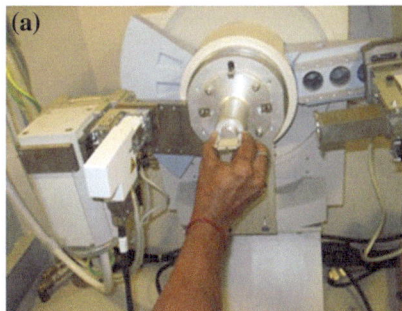
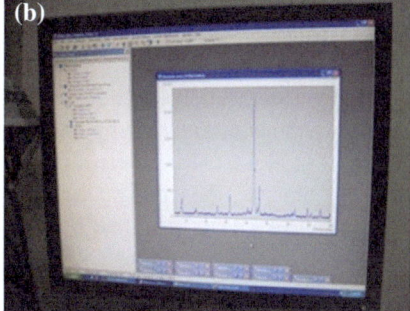

Fig. 4.3 Sediment added to the machine (**a**) and generation of diffractogram (**b**)

Fig. 4.4 Methods of mineral identification from diffractogram

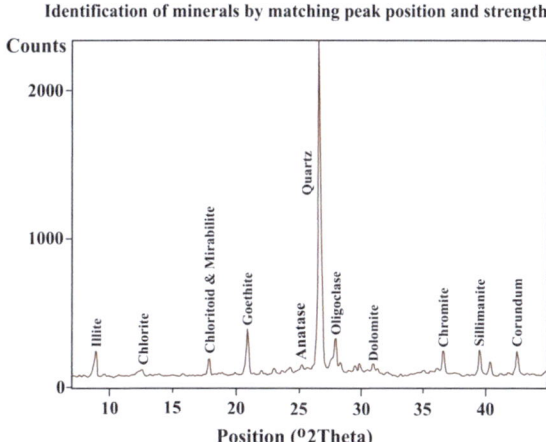

conditions of the diffractometer under which the sediments were analyzed are given in Table 4.3. Different minerals were identified from the diffractograms by analyzing the peak's position, intensity, shape and breadth (Klug and Alexander 1954) (Fig. 4.4). Peak position is identified using *Bragg's formula* which is expressed as

Table 4.3 Instrumental conditions of XPERT-PRO diffractometer (Maity and Maiti 2016)

1. Wavelength: 1.5406 Å Cu Kα
2. Generator power: 40 kV and 30 mA
3. Goniometer type: θ/θ
4. Goniometer radius: 240 mm
5. Emitting slits: 2 and 4 mm
6. Receiving slits: 0.5 and 0.3 mm
7. Scan type: Continuous scan
8. Goniometer speed: 1°/min
9. Acquisition step size: 0.03°
10. Scanned interval: 7°–45°2θ

$n\lambda = 2d \sin \theta$ (n = a whole number, λ = X-ray wavelength, d = the distance between planes of atoms and θ = the angle of incidence). Most of the important peaks for mineral recognition are detected between the values of 8° and 43°2θ. The qualitative classification of minerals was done by searching the strongest peak or peaks for a minerals and then by finding the positions of weaker peaks for the same mineral (Fig. 4.4). Once a series of peaks was verified as the finest match for a particular mineral, then the weaker peaks were removed from consideration. All remaining peaks were identified by the repetition of the same technique.

4.4.3 Quantification of Different Minerals in Sediments

The quantification (%) of different minerals present in the sediment samples is done using the *'matrix flushing method'* as proposed by Chung (1974). This method is independent of the matrix effect; all absorption aspects are precisely flushed out. It represents a perfect relationship between the intensity and concentration as no assumption or approximation is considered. The functioning Eq. (4.1) of this matrix flushing method is very easy to understand, no complicated computations are implicated.

$$X_i \left(\frac{X_c}{k_i}\right)\left(\frac{I_i}{I_c}\right) \tag{4.1}$$

X_i = weight fraction of component mineral i, X_c = weight fraction of corundum (*the flushing agent*), k_i = the Reference Intensity Ratio of mineral i, I_i the intensity of X-rays diffracted by a selected plane (hkl) of mineral i, and I_c the intensity of X-rays diffracted by a selected plane (hkl) of corundum (*flushing agent*). Corundum (α-Al_2O_3) has been chosen for Reference Intensities by the Powder Diffraction File as it is pure, stable and available. In my study, corundum is used as the flushing agent for the same justification. The Reference Intensities of different minerals were detected using the same diffractometer under the same instrumental configurations. Sediment samples were treated and prepared suitably for the quantification of various minerals by matrix flushing technique. A fixed quantity of corundum (flushing agent) was then added into the sediment samples. All the peaks in the *diffractogram* were accurately identified, and the intensity of the strongest line of each mineral component was quantified under the same instrumental configurations. The amount of minerals (in percentage) in sediment samples was calculated following the Eq. (4.1) and is shown in Tables 4.4 and 4.5.

4.4 Materials and Methodology of the Study

Table 4.4 Mineralogy of sediment samples collected from river bed (Maity and Maiti 2016)

Location	% amount of minerals								
	Quartz (qz)	Oligoclase (ol)	Dolomite (do)	Tourmaline (tr)	Calcite and staurolite (cs)	Dolomite and epidote (de)	Ilmenite and actinolite (ia)	Garnet (pyrope) (gp)	Chromite (cr)
Bandar (S1)	61.6	5.02	2.38	2.91	0	0	0	0	4.90
Jasar (S2)	58.79	4.82	0	2.28	0	2.54	0	0	4.26
Kolaghat (S3)	60.51	7.57	2.01	0	0	0	0	0	3.29
Kantapukur (S4)	65.36	9.88	0	0	0	0	0	2.41	5.80
Gopinathpur (S5)	68.49	2.95	0	0	2.07	2.08	0	2.06	3.07
Soyadighi (S6)	58.27	11.39	0	0	0	0	0	0	2.33
Telipara (S7)	63.19	10.1	0	0	2.03	2.49	0	0	4.51
Anantapur (S8)	62	4.35	0	0	0	4.31	2.61	0	7.37
Pyratungi (S9)	62.08	9.32	2.11	0	2.11	0	0	0	2.51
Amberia (S10)	65.29	5.01	0	0	2.07	0	0	0	4.02
Dhanipur (S11)	55.07	9.08	2.07	0	2.02	0	0	0	2.59

(continued)

Table 4.4 (continued)

Location	% amount of minerals							
	Quartz (qz)	Oligoclase (ol)	Dolomite (do)	Tourmaline (tr)	Calcite and staurolite (cs)	Dolomite and epidote (de)	Ilmenite and actinolite (ia)	Garnet (pyrope) (gp)	Chromite (cr)
Bholsara (S12)	59.40	5.10	0	0	0	0	0	2.08	3.10
Geonkhali (S13)	59.15	4.37	0	0	0	0	0	2.09	3.22

Location	% amount of minerals								
	Sillimanite (si)	Corundum (co)	Illite (il)	Chlorite (ch)	Chloritoid and mirabilite (cm)	Goethite (go)	Anatase (an)	Laumontite and strontianite (ls)
Bandar (S1)	3.28	10.18	5.79	2.69	3.38	10	2.78	0
Jasar (S2)	3.54	3.65	5.33	2.3	2.53	7.91	2.02	0
Kolaghat (S3)	4.05	4.32	4.39	2.01	2.37	7.49	2.02	0
Kantapukur (S4)	3.52	6.37	2.68	2.03	2.15	9.31	0	0
Gopinathpur (S5)	3.03	3.77	3.84	2.04	2.73	8.01	0	0
Soyadighi (S6)	2.86	3.94	7.21	2.15	4.22	7.62	0	0
Telipara (S7)	3.56	4.77	5.09	2.04	2.25	10.06	0	0
Anantapur (S8)	2.29	6.01	2.11	2.06	2.09	7.95	0	0
Pyratungi (S9)	3.83	4.34	7.01	2.83	4.15	9.61	0	0
Amberia (S10)	3.18	4.04	3.03	2.05	2.17	7.05	0	2.03
Dhanipur (S11)	2.92	2.99	9.4	2.75	5.58	5.53	0	0
Bholsara (S12)	2.65	3.15	9.80	3.59	4.77	6.35	0	0
Geonkhali (S13)	2.69	3.01	10.24	3.79	5.37	6.21	0	0

4.5 Mineralogy of Sediments

4.5.1 Mineralogy of River Bank Sediment Samples

The mineralogical composition of sediments of both the river banks is almost alike. All the sediments are dominated mainly by illite (24–32%) and smectite (25–31.3%) minerals (Table 4.5). Chlorite (10.22–13%), kaolinite (8.12–12%) and quartz (7.5–9.34%) are also important constituent minerals detected in sediment samples. Goethite (3–6.13%), feldspar (2.2–6%) and mixed layer minerals (1.6–5%) are present in very small proportion (Table 4.5). Mixed layer mineral is not detected in sediment sample 1 in the left bank and sample 7 in the right bank. Laumontite and strontianite (4.15%) is detected only in the sediment sample 3 in the left bank (Table 4.5).

4.5.2 Mineralogy of River Bed Sediment Samples

Quartz is the mineral having strongest peak in diffractogram in all sediment samples, along with illite, chlorite, chloritoid, anatase, goethite, oligoclase, sillimanite and corundum. The proportion of quartz is greater than 55% in all the sediment samples. The highest amount (68.49%) of quartz is measured in the sample S5; whereas lowest amount (55.07%) is observed in the sample S11 (Fig. 4.5 and Table 4.4). Quartz (55.07–68.49%), illite (2.11–10.04%), chlorite (2.01–3.79%), chloritoid and mirabilite (2.15–5.58%), oligoclase (2.95–11.39%), goethite (5.53–10.06%), chromite (2.33–7.37%), sillimanite (2.29–4.05%) and corundum (2.99–10.18%) are the minerals, detected in all the sediment samples (Fig. 4.5 and Table 4.4). Tourmaline is observed in the sediment samples S1 and S2 only; while anatase is detected in the samples S1, S2 and S3. Laumontite and strontianite are identified in very little proportion (2.03%) only in sediment sample S10 (Fig. 4.5 and Table 4.4). Calcite and staurolite are detected in the sediment samples S5, S7, S9, S10 and S11; whereas ilmenite and actinolite (2.61%) are identified in the sample S8 only. Garnet (Pyrope), the unique mineral, is observed in the sediment samples S4, S5, S8 and S12. Dolomite and epidote are identified in the sediment samples S2, S5, S7 and S8 (Fig. 4.5 and Table 4.4).

Chi-square (x^2 test) statistical test signifies that quartz mineral shows steady trend in distribution in all the sediment samples. At 11 degree of freedom, at 5% significance level the estimated value of x^2 is 2.7162392 but the critical value is 19.68, which reveals that the variation of spatial distribution of percentage of quartz in sediment samples is insignificant (Table 4.6). Similar result is found for all other minerals except tourmaline and anatase. At 5% significance level the spatial variation of percentage of tourmaline and anatase is significant, which designates the unequal distribution of these two minerals in the lower reach (Table 4.6).

Table 4.5 Mineralogy of sediment samples collected from river banks (Maity and Maiti 2016)

Sediment sample	% amount of minerals								
	Kaolinite	Chlorite	Smectite	Illite	Quartz	Laumontite and strontianite	Goethite	Feldspar	Mixed layer mineral
1 (left bank)	10.52	11.23	28	30.6	9.34	0	6.11	2.2	0
2 (left bank)	11.5	10.5	27.31	24	9.18	0	5	4.1	4.26
3 (left bank)	8.12	10.22	31.3	32	8.11	4.15	4.15	3.1	3
4 (left bank)	10.6	11	29.23	29.17	8	0	3	6	3
5 (right bank)	11.23	12	27	30.5	7.5	0	5.17	5	1.6
6 (right bank)	11.2	10.23	25	29	9.32	0	6.13	4.12	5
7 (right bank)	12	10.64	31	30.20	8	0	4.16	3	0
8 (right bank)	10.5	13	29	31.2	7.63	0	3.17	2.5	3

4.6 Principal Component Analysis (PCA) of Bed Sediment Minerals

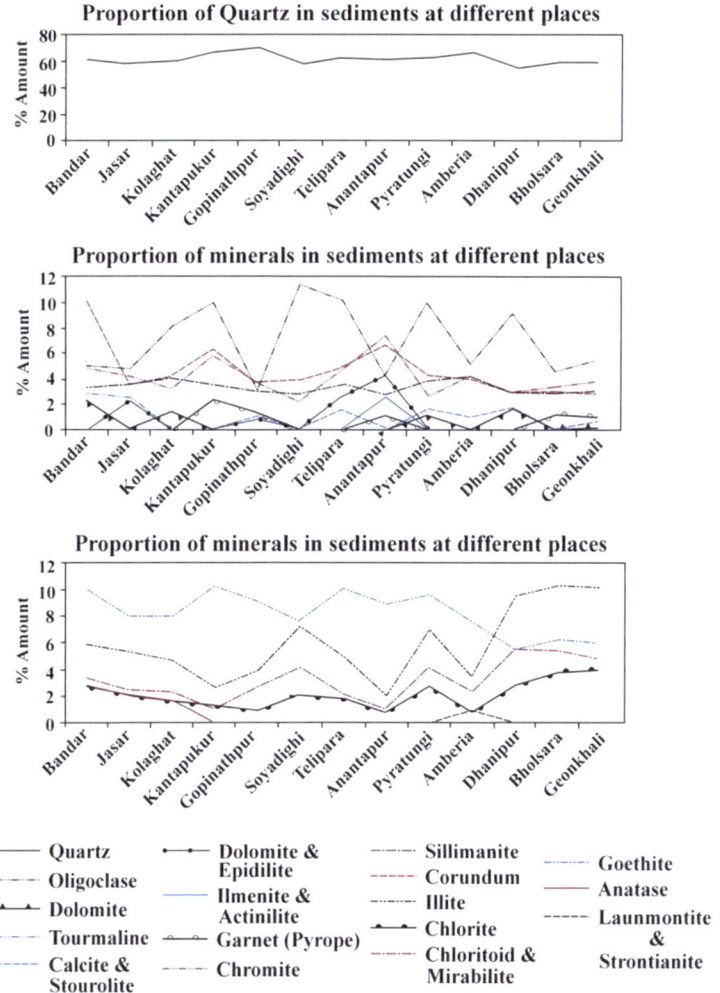

Fig. 4.5 Quantity of different minerals in sediments (Maity and Maiti 2016). *Source* Laboratory experiment

4.6 Principal Component Analysis (PCA) of Bed Sediment Minerals

Principal component analysis (*PCA*) is used as a standard multivariate statistical technique mainly for its two advantages: (1) Reduction of the number of correlated variables and to extract smaller number of uncorrelated principal components which signify the majority of the variability contributed by the multiple variables and (2) to improve the interpretability of the components as combinations of multiple

Table 4.6 Statistical significance of minerals distribution (Maity and Maiti 2016)

Minerals	Calculated value of x^2	Degree of freedom	Critical value of x^2 (5% significance level)	Remarks
Quartz	2.7162392	11	19.68	Equal distribution at different locations
Oligoclase	11.652523			Equal distribution at different locations
Dolomite	15.824156			Equal distribution at different locations
Tourmaline	*25.976346*			*Unequal distribution at different locations*
Calcite and staurolite	9.7855428			Equal distribution at different locations
Dolomite and epidote	14.527632			Equal distribution at different locations
Ilmenite and actinolite	2.4364315			Equal distribution at different locations
Garnet pyrope	16.236289			Equal distribution at different locations
Chromite	2.8253219			Equal distribution at different locations
Sillimanite	0.5197432			Equal distribution at different locations
Corundum	8.5853876			Equal distribution at different locations
Illite	9.875198			Equal distribution at different locations
Chlorite	3.8962987			Equal distribution at different locations
Chloritoid and mirabilite	6.2649865			Equal distribution at different locations
Goethite	2.5358732			Equal distribution at different locations
Anatase	*20.022946*			*Unequal distribution at different locations*
Laumontite and strontianite	10.152502			Equal distribution at different locations

variables (Cheng et al. 2006). PCA measures the interrelationship among multiple variables with the help of correlation (covariance) matrix.

Total seventeen variables (minerals), detected from XRD analysis of 13 river bed sediment samples and their percentage value were used for Principal Component Analysis in this work. These minerals are quartz (qz), oligoclase (ol), dolomite (do),

4.6 Principal Component Analysis (PCA) of Bed Sediment Minerals

tourmaline (tr), calcite and staurolite (cs), dolomite and epidote (de), ilmenite and actinolite (ia), garnet pyrope (gp), chromite (cr), sillimanite (si), corundum (co), illite (il), chlorite (ch), chloritoid and mirabilite (cm), goethite (go), anatase (an), laumontite and strontianite (ls). The *matrix of inter correlations* (R) among the percentage amount of these minerals is calculated and is shown in Table 4.7. The nature of bi-variate relationship among the percentage amount of these minerals in sediment samples is easily understood from the correlation matrix.

The Eigen values of the matrix (R) have been calculated and these 17 Eigen values in descending order are 5.071, 3.357, 2.686, 1.624, 1.399, 0.931, 0.851, 0.553, 0.293, 0.180, 0.050, 0.005, 4.058×10^{-15}, 2.668×10^{-15}, -9.137×10^{-17}, -4.31×10^{-15} and -1.070×10^{-14}. There are five Eigen values clearly greater than unity as 5.071, 3.357, 2.686, 1.624 and 1.399. These five Eigen values contribute for $\frac{5.071 + 3.357 + 2.686 + 1.624 + 1.399}{17} \times 100 = 83.154\%$ of the total variation of the data matrix (Table 4.8).

The PCA1 contribute for about 29.828% of the total variation and is dominated by quartz, chromite and Goethite, which have positive loadings and illite, chlorite, chloritoid and mirabilite, having negative loadings (Table 4.8), i.e., this component is more significant where these minerals don't present high values. The PCA2 contributes for the 19.748% of the total variation of the data. It is represented by dolomite, tourmaline, sillimanite and anatase, all of them with positive loadings (Table 4.8), i.e., this component is more important where these minerals do present high values. The PCA3 accounts for about 15.798% of the total variation of the data and is dominated by calcite and staurolite having negative loading (Table 4.8). The PCA4 and PCA5 jointly contribute for 17.780% of the total variation of the mineral distribution (Table 4.8). In that case, Garnet Pyrope has positive loading whereas Laumontite and Strontianite have negative loading.

4.7 Understanding the Sources of Sediments

The sediments of lower reach of the Rupnarayan River is dominated by quartz mineral with minor presence of goethite, illite, chlorite, chloritoid and mirabilite, chromite, oligoclase, sillimanite and corundum (Fig. 4.5 and Table 4.4). The proportion of tourmaline in the rocks of the upper catchment is very little and it is detected in the sediments at Bandar (S1) and Jasar (S2) only; while anatase mineral is distributed up to Kolaghat (S3) from the upper catchment area (Fig. 4.5 and Table 4.4). Though, laumontite and strontianite is not detected in the rocks of upper catchment but it has been detected in very little proportion (4.15%) in the sediment sample 3 on the left bank of the lower reach. The presence of laumontite and strontianite in the sediment at Amberia (S10) indicates that these minerals have been supplied to the river from the local river bank source. The specific gravity of different minerals play significant role to control the distribution of these minerals in the lower reach. Because of high specific gravity (2.83–3.32 and 3.90 respectively), tourmaline

Table 4.7 Co-relation matrix of different minerals

	qz	ol	do	tr	cs	de	ia	gp	cr	si	co	il	ch	cm	go	an	ls
qz	1	-.267	-.310	-.138	.320	.230	.044	.329	.329	.184	.269	-.707	-.471	-.673	.505	-.163	.322
ol		1	.202	-.301	.129	-.325	-.268	-.272	-.232	.383	-.039	.105	-.235	.112	.259	-.227	-.197
do			1	.288	.143	-.417	-.192	-.442	-.259	.443	.344	.201	.105	.266	.144	.493	-.192
tr				1	-.334	.072	-.122	-.281	.211	.182	.597	-.041	.023	-.122	.345	.835	-.122
cs					1	.017	-.228	-.198	-.330	.198	-.282	-.053	-.193	.004	.074	-.426	.368
de						1	.707	-.181	.605	-.264	.050	-.489	-.407	-.518	.214	-.049	-.181
ia							1	-.192	.712	-.530	.208	-.410	-.209	-.296	.004	-.156	-.083
gp								1	-.022	-.261	-.175	.167	.392	.177	-.185	-.358	-.192
cr									1	-.191	.589	-.671	-.368	-.642	.385	.121	.020
si										1	.127	-.243	-.349	-.360	.511	.443	-.003
co											1	-.418	-.236	-.373	.685	.531	-.095
il												1	.859	.949	-.499	-.123	-.309
ch													1	.816	-.421	-.104	-.214
cm														1	-.549	-.233	-.278
go															1	.270	-.180
an																1	-.156
ls																	1

4.7 Understanding the Sources of Sediments

Table 4.8 Depiction of loadings on principal components

Component loading	PCA1	PCA2	PCA3	PCA4	PCA5
Quartz (qz)	**.628**	−.295	−.397	.522	.078
Oligoclase (ol)	−.147	.292	−.486	−.476	**.538**
Dolomite (do)	−.140	**.776**	−.045	−.219	−.108
Tourmaline (tr)	.283	**.647**	.483	.214	−.253
Calcite and staurolite (cs)	−.046	−.128	**−.734**	−.133	−.245
Dolomite and epidote (de)	.598	−.428	.298	−.334	−.006
Ilmenite and actinolite (ia)	.458	−.469	.490	−.472	−.015
Garnet pyrope (gp)	−.251	−.434	.105	**.713**	.423
Chromite (cr)	**.778**	−.224	.419	−.051	.110
Sillimanite (si)	.230	**.659**	−.553	.102	.105
Corundum (co)	.610	.476	.313	.149	.145
Illite (il)	**−.936**	.150	.206	−.020	.053
Chlorite (ch)	**−.766**	.011	.413	.260	.021
Chloritoid and mirabilite (cm)	**−.934**	−.078	.197	−.080	.045
Goethite (go)	**.668**	.401	−.172	.089	.425
Anatase (an)	.294	**.777**	.366	.149	−.258
Laumontite and strontianite (ls)	.150	−.230	−.427	.177	**−.702**
Eigen value	5.071	3.357	2.686	1.624	1.399
% of variance	29.828	19.748	15.798	9.552	8.228
Cumulative % of variance	29.828	49.576	65.374	74.926	83.154

and anatase have been transported comparatively less distance (up to Jasar [S2] and Kolaghat [S3] respectively) from the upper catchment area.

In contrast to this, due to low specific gravity, quartz (2.65) has been transported a long distance towards downstream and is detected in all the sediment samples with very high proportion (Figs. 4.5, 4.6a and Table 4.4). Specific gravity is also low for the minerals like, calcite (2.69–2.71), oligoclase (2.64–2.66), chlorite (2.54–2.78), illite (2.6–2.9) and dolomite (2.8–2.9) and these minerals are identified in all the sediment samples (Figs. 4.5, 4.6b, c, 4.7d and Table 4.4). The change of forms of some minerals into other forms is very significant to explain the distribution of minerals in the lower reach. Some of the minerals change their form easily than the other minerals. Quartz is very rigid and is not easily dissolved, thus it prevents itself from both physical and chemical weathering. Because of this, it is the common mineral in all the sediment samples in the lower reach.

The upper and middle catchment area of the Rupnarayan River is mostly composed of granite and gneiss (Fig. 1.3 in Maity and Maiti 2018) and becomes the source of supply of quartz as the dominant constituent mineral in all the sediment samples in the lower reach. Minerals like feldspar, plagioclase, augite, olivine and other pyroxene minerals are very susceptible to chemical reactions and form new minerals quickly. Because of this these minerals are substantially less common in sediment samples. Feldspar, plagioclase, olivine and pyroxene minerals are easily

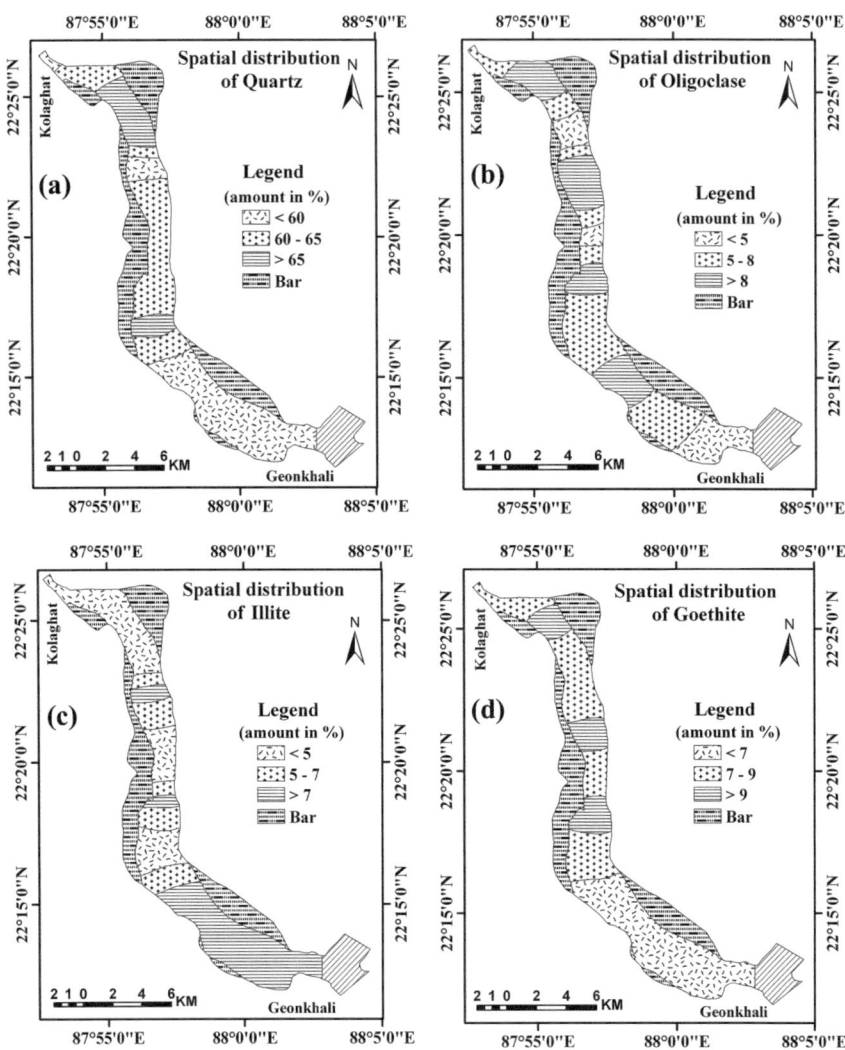

Fig. 4.6 Spatial distribution of quartz (**a**), oligoclase (**b**), illite (**c**) and goethite (**d**) in the lower reach

converted into clay minerals and produce illite, smectite and chlorite (Loughnan 1969; Eggleton and Boland 1982; Huang 1989). Epidote group minerals are produced due to easy alteration of plagioclase feldspar. Pyroxene and amphibole minerals are easily converted into chlorite group minerals (Churchman 1980). Chlorite, epidote and actinolite are produced due to quick alteration of augite, a significant pyroxene group rock forming mineral (Allen and Hajek 1989). Due to this, minerals like feldspar, plagioclase, olivine, augite, amphibole and pyroxene are not detected in the sediments in the lower reach though they are the significant rock

4.7 Understanding the Sources of Sediments

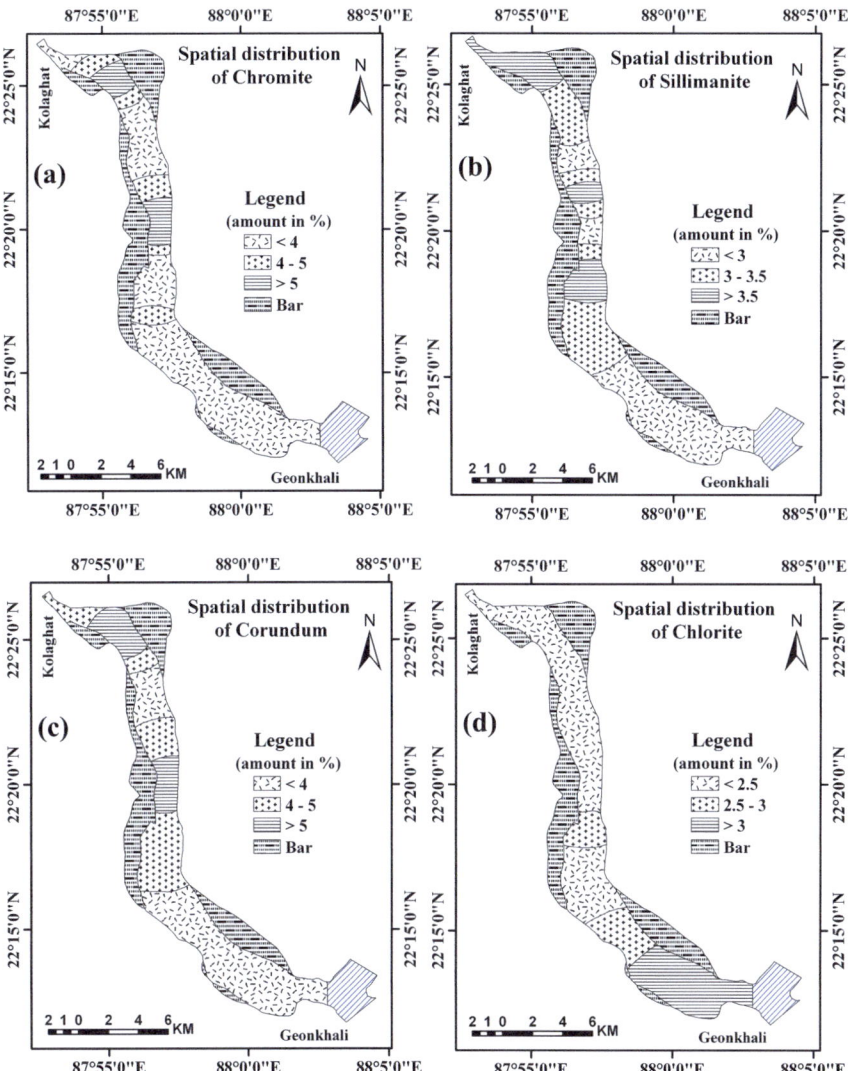

Fig. 4.7 Spatial distribution of chromite (**a**), sillimanite (**b**), corundum (**c**) and chlorite (**d**) in the lower reach

forming minerals in the upper and middle catchment area of the studied river (Maity and Maiti 2016).

No noticeable trend in the spatial distribution of the different minerals either towards upstream or downstream is observed in the lower reach. Spatial distribution of the minerals reveals that the distribution of minerals is unsystematic in the lower reach (Figs. 4.6, 4.7 and 4.8). The distribution of a small number of minerals can be explained systematically. The high percentage of illite in some sediment samples is

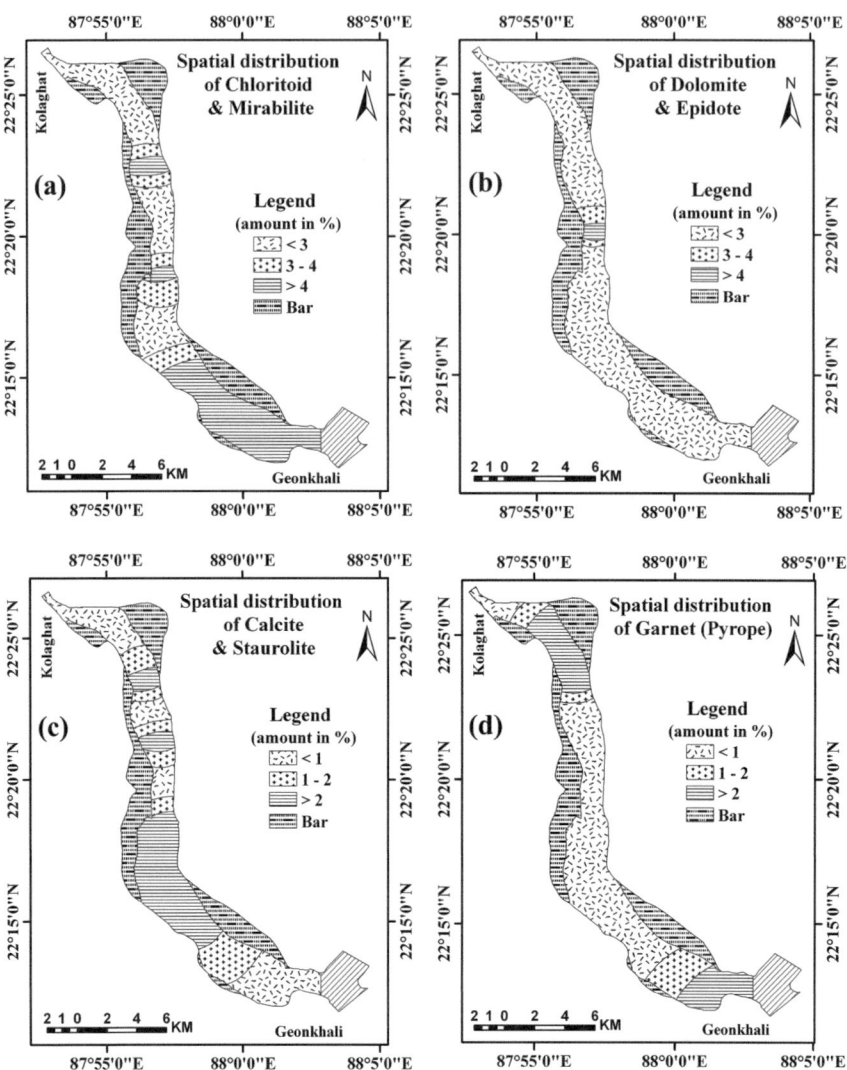

Fig. 4.8 Spatial distribution of chloritoid and mirabilite (**a**), dolomite and epidote (**b**), calcite and staurolite (**c**) and garnet pyrope (**d**) in the lower reach

associated with the more concentration of chlorite, chloritoid and mirabilite (Figs. 4.5, 4.6c, 4.7d, 4.8a and Table 4.4). But the percentage amount of minerals like, goethite, chromite and corundum is less in those samples (Figs. 4.6d, 4.7a, c and Table 4.4). The samples within which the percentage of oligoclase is high, the percentage of dolomite and epidote are measured to be less (Figs. 4.6b and 4.8b). But the distribution of other minerals is found to be unsystematic and hapazard in the lower reach (Figs. 4.6, 4.7 and 4.8) (Maity and Maiti 2016).

4.7 Understanding the Sources of Sediments

In the lower reach of the studied river, having *estuarine characteristics*, there is a mixing of sediments and minerals in both the tidal phases (during high and low tide). Minerals which are drained from the upper and middle catchment are caught up in the estuary and are redistributed again towards upstream by stronger flood tide. This mechanism is responsible for such unsystematic and irregular distribution of minerals in the lower reach (Maity and Maiti 2016). Thus, it can be concluded that the upper and middle parts of the catchment area are the main source of minerals as well as sediments deposited in the *lower reach* of the *Rupnarayan River*. Some of the minerals including quartz, illite, chlorite and goethite are also supplied to the river from the lower catchment area because these minerals have been identified in the sediments of the lower catchment area. Erosion of river banks is also an important source of sediments in the lower reach of the river, because quartz, illite, chlorite, goethite, laumontite and strontianite have been identified in the sediment samples collected from the river banks.

References

Allen BL, Hajek BF (1989) Mineral occurrence in soil environments. In: Dixon JB, Weed SB (eds) Minerals in soil environments, 2nd edn. Soil Science Society of America, Madison, pp 199–278

Blatt H, Middleton G, Murray R (1980) Origin of sedimentary rocks. Prentice-Hall, Englewood Cliffs

Chase WT (1971) Egyptian blue as a pigment and ceramic material. In: Brill RH (ed) Science and archaeology. Massachusetts Institute of Technology, Boston, pp 80–90

Chen PY (1977) Table of key lines in X-ray powder diffraction patterns of minerals in clays and associated rocks. Indiana Geol Survey Occas Pap 21:1–67

Cheng Q, Jing L, Panahi A (2006) Principal component analysis with optimum order sample correlation coefficient for image enhancement. Int J Remote Sens 27(16):3387–3401

Chung FH (1974) Quantitative interpretation of X-ray diffraction patterns of mixtures. I. matrix-flushing method for quantitative multicomponent analysis. J Appl Cryst 7:519–525

Churchman GJ (1980) Clay minerals formed from micas and chlorites in some New Zealand soils. Clay Miner 15:59–76

Eggleton RA, Boland JN (1982) Weathering of enstatite to talc through a sequence of transitional phases. Clays Clay Miner 30:11–20

FitzHugh EW, Gettens RJ (1971) Calcite and other efflorescent salts on objects stored in wooden museum cases. In: Brill RH (ed) Science and archaeology. Massachusetts Institute of Technology, Boston, pp 91–105

Friedman GM (1961) Distinction between dune, beach and river sands from their textural characteristics. J Sediment Petrol 31(4):514–529

Ghosh SK, Datta NP (1974) X-ray investigation of clay minerals in the soils of West Bengal. Proc Indian Sci Acad 40:138–150

Gibbs RJ (1965) Quantitative X-ray diffraction analysis using clay mineral standards extracted from the samples to be analyzed. Am Mineral 50:741

Gibbs RJ (1967) Quantitative X-ray diffraction analysis using clay mineral standards extracted from the samples to be analysed. Clay Miner 7:79–90

Huang PM (1989) Feldspars, olivines, pyroxenes, and amphiboles. In: Dixon JB, Weed SB (eds) Minerals in soil environments, 2nd edn. Soil Science Society of America, Madison, pp 975–1050

Klug HP, Alexander LE (1954) X-ray diffraction procedures for polycrystalline and amorphous materials. Wiley, New York

Loughnan FC (1969) Chemical weathering of the silicate minerals. Elsevier, New York

Maity SK, Maiti RK (2016) Understanding the Sources of sediments from mineral composition at lower reach of the Rupnarayan River, West Bengal, India—an x-ray diffraction (XRD) based analysis. GeoResJ 9(12):91–103

Maity SK, Maiti RK (2018) Sedimentation in the Rupnarayan River: hydrodynamic processes under a tidal system. Springer briefs in earth sciences. Springer, Berlin

Matson FR (1971) A study of temperature used in firing ancient Mesopotamian pottery. In: Brill RH (ed) Science and archaeology. Massachusetts Institute of Technology, Boston, pp 65–79

Mukhopadhyay SC, Dasgupta A (2010) River dynamics of West Bengal, physical aspects, vol 1. pp 1–220

O'Malley LSS (1995) Physical aspects, Bengal District Gazetteers, Midnapore. Published by West Bengal District Gazetteers. pp 1–141

Rex RW, Bauer WR (1965) New amine reagents for X-ray determination of expandable clays in dry samples. In: Proceedings of the 13th national conference on clays and clay minerals, pp 411–418

Rex RW, Chown RG (1960) Planchet press and accessories for mounting X-ray powder diffraction samples. Am Mineral 45:1280–1282

Rittenhouse G (1943) A visual method of estimating two dimensional sphericity. J Sediment Petrol 13:79–81

Shepard AO (1971) Ceramic analysis: the implications of methods; the relations of analysts and archaeologists. In: Brill RH (ed) Science and archaeology. Massachusetts Institute of Technology, Boston, pp 55–64

Tankersley KB (1995) Paleoindian contexts and artifact distribution patterns at the Bostrom Site, St. Clair County, Illinois. MidCont J Archaeol 20:40–61

Tankersley KB, Balantyne MR (2010) X-ray powder diffraction analysis of Late Holocene reservoir sediments. J Archaeol Sci 37:133–138

Tankersley KB, Meinhart J (1982) Physical and structural properties of ceramic materials utilized by a Fort Ancient Group. MidCont J Archaeol 7:225–273

Tankersley KB, Bassett J, Frushour S (1985) A gourd bowl from salts cave, Kentucky. Tenn Anthropologist 10:95–104

Tankersley KB, Munson CA, Munson PJ et al (1990) The mineralogy of Wyandotte Cave aragonite, Indiana, and its archaeological significance. In: Lasca NP, Donahue J (eds) Archaeological geology of North America. Geological Society of America, Boulder, pp 219–230

Tankersley KB, Shaffer N, Hess M et al (1995) They have a rock that bleeds: sunrise red ochre and its early paleoindian occurrence at the Hell Gap site, Wyoming. Plains, Anthropologist 40:185–195

Trachsel M, Eggenberger U, Grosjean M et al (2008) Mineralogy based quantitative precipitation and temperature reconstructions from annually laminated lake sediments (Swiss Alps) since AD 1580. Geophys Res Lett 35:L13707

Zong P, Becker RT, Ma XP (2015) Upper Devonian (Famennian) and Lower Carboniferous (Tournaisian) ammonoids from western Junggar, Xinjiang, northwestern China—Stratigraphy, taxonomy and palaeobiogeography. Palaeobio Palaeoenv 95:159–202

Chapter 5
Conclusion

Abstract In the lower reach of the Rupnarayan River, *sedimentation* is the result of the combined interaction between *fluvial and marine processes*. Seasonal variation of *available shear stress* during high and low tide in connection to *critical shear stress* is the main factor controlling *sedimentation*. Sediments are mostly transported by suspension with rolling and graded suspension and are deposited in the environment with low to moderately low energy conditions. Sediments are mainly supplied from the *upper and middle catchment* of the river with little contribution from the river banks. The result of the work will act as a decision supporting system to the engineers, hydrologists, planners and other concerned authorities, working on the aspects of *sedimentation* and management of *associated problems*.

Keywords Fluvial and marine processes · Available and critical shear stress Depositional environment · Sources of sediments

Detailed studies on *causes, mechanisms* and *magnitude* of sedimentation reveal that, in the *lower reach* of the *Rupnarayan River*, sedimentation is the result of the combined interaction between fluvial and marine processes. The seasonal variation of available shear stress during low tide and high tide in connection to critical shear stress (related to grain size of sediment) is the main factor controlling sedimentation. The *available shear stress*, in most of the cases during low tide (except monsoon season) is lower than the critical shear stress required for entrainment of sediment and this is found to be the main reason of sedimentation in the area under study, especially in non-monsoon season when this deviation becomes maximum. Most of the places, having deficiency of energy (available shear stress is lower than critical shear stress), during low tide are characterized by deposition of sediments. Sedimentation in some places is affected by the factors like sheltering, imbrications, packing of grains, grain-fabric effects, adhesion forces and organic mats. Very fine grained sand, silt and clay particles are sheltered by moderately coarse and coarse sand particles, which increase the critical shear stress of entrainment and restrict the sediments to be eroded. The presence of mud (silt and clay) above the critical limit (30%) in some of the sediment samples (14 samples in pre-monsoon, 6 samples in monsoon and 11 samples in post-monsoon) generates the *cohesive property* and restricts the entrainment of sediments. The proportion of mud is more in dry season which increases the cohesive property, restricts the entrainment of sediments and increases the rate of sedimentation. During prolonged dry period (scarcity of terrestrial discharge) the growth of *surface mat of algae* on river bed creates a binding force between clay particles in muddy sediments which increases the critical shear stress of sediment entrainment.

Sedimentation, in the area under study is taking place in *estuarine environment* and most of the sediments are transported by suspension with rolling and graded suspension. The dominance of very fine sand (63.80%) and coarse silt (24.44%) type of sediments indicates low and moderately low energy conditions of the depositional environment. Kurtosis values indicate the *high energy environment* in monsoon season than that of in non-monsoon season. During dry season, the marine environment becomes dominant over the fluvial environment due to insufficient riverine discharge, but during monsoon season, occurrence of huge rainfall increases the riverine discharge and reduces the influence of marine processes. *XRD* analysis reveals that *quartz* is the main mineral in all the bed sediment samples (>55%) along with illite, chlorite, chloritoid, goethite, oligoclase, corundum, chromite etc. The minerals identified in bed sediments have their origin in the upper and middle catchment which indicates that most of the sediments at the lower reach are drained from the *upper and middle catchment*. Some of the minerals (illite, chlorite, quartz, goethite, laumontite and strontianite) in little proportion are also supplied from the lower part of the catchment area and from the erosion of the *river banks*.

Most of the earlier works, in the study area, mainly by the Hydrologists and Engineers are based on short duration monitoring and location specific. In the present study attempt has been made to understand the process of interaction between marine and fluvial processes in a holistic approach that leads to sedimentation. All sorts of hydrological, bathymetric and geometric characteristics of the river, responsible for sedimentation have been studied and monitored in detail. The geological, geomorphological and hydrological characteristics of the upper and middle catchments of the river have also been studied to understand and explain the characters at the source region. The study has been linked and compared with mostly accepted works of different researchers. In most of the cases the result of the study agreed with the previous works. The result of the work will act as a *decision supporting system* to the engineers, hydrologists, planners and other concerned authorities, working on the aspects of sedimentation and *management of associated problems* not only in case of the area under study but also in any of the river in the world, having similar climatic, geological, geomorphological, hydrological, bathymetric and hydraulic characteristics.

Index

A
Adhesion force, 36
Available shear stress, 2, 3, 5–35

B
Bi-variate plot, 50, 52
Bragg's Law, 63

C
Chi-Square test, 67
C-M plotting, 40, 41
Cohesive property of sediment, 36
Co-relation matrix, 72
Critical shear stress, 2, 3, 5–35, 37

D
Degree of freedom, 67, 70
Diffractogram, 62–64, 67

E
Estuarine environment, 79

F
Flushing agent, 64
Friction, 35

G
Gradient of river bed, 6, 7, 22
Grain-fabric effects, 35
Grain sheltering, 5

H
High and low tide, 2, 6, 7, 20, 26, 34, 77
Hoogly estuary, 47

I
Hydrodynamic processes, 40, 41, 54

Initiation of motion, 6, 35, 37

K
Kurtosis, 3, 40, 79

L
Linear Discriminate Analysis (LDA), 3, 40, 41

M
Matrix flushing method, 64
Mineralogical composition, 58, 62, 67

N
Non-uniform sediment, 35, 36

O
Organic mat, 7

P
Packing of grains, 7, 35
Principal Component Analysis (PCA), 69, 70

R
River discharge, 51

S
Sedimentation, 2, 3, 6, 7, 9, 13, 20, 26, 33–37, 47, 48, 51, 53, 58, 59, 80
Sediment discharge, 22, 39, 41, 43, 45, 46, 48, 80
Sediment entrainment, 30, 35–37, 79

Shallow marine environment, 41–46
Shoal area, 33
Sieving technique, 7, 40
Skewness, 3, 40, 41, 51, 52
Sorting of grains, 3, 35
Stream energy, 1, 2, 33, 34, 48

T
Tidal asymmetry, 2, 47, 58
Trapping of sediment, 34, 58, 59, 71, 77, 80

W
Water discharge, 2
Water velocity, 1, 3, 6, 22, 26, 34

X
X-Ray Diffraction (XRD), 57–59, 61, 62, 70, 80

If you have any concerns about our products,
you can contact us on
ProductSafety@springernature.com

In case Publisher is established outside the EU,
the EU authorized representative is:
Springer Nature Customer Service Center GmbH
Europaplatz 3, 69115 Heidelberg, Germany

Printed by Libri Plureos GmbH
in Hamburg, Germany